MEDIA RUINS

Labor and Technology

Winifred Poster, series editor

Madison Van Oort, *Worn Out: How Retailers Surveil and Exploit Workers in the Digital Age and How Workers Are Fighting Back*

Sofya Aptekar, *The Green Card Soldier: Between Model Immigrant and Security Threat*

Margaret Jack, *Media Ruins: Cambodian Postwar Media Reconstruction and the Geopolitics of Technology*

MEDIA RUINS
CAMBODIAN POSTWAR MEDIA RECONSTRUCTION AND THE GEOPOLITICS OF TECHNOLOGY

Margaret Jack

THE MIT PRESS
CAMBRIDGE, MASSACHUSETTS
LONDON, ENGLAND

© 2023 Massachusetts Institute of Technology

This work is subject to a Creative Commons CC BY-NC-ND license. Subject to such license, all rights are reserved.

The MIT Press would like to thank the anonymous peer reviewers who provided comments on drafts of this book. The generous work of academic experts is essential for establishing the authority and quality of our publications. We acknowledge with gratitude the contributions of these otherwise uncredited readers.

This book was set in ITC Stone and Futura Std by New Best-set Typesetters Ltd. Printed and bound in the United States of America.

Library of Congress Cataloging-in-Publication Data

Names: Jack, Margaret (Margaret Cora), author.
Title: Media ruins : Cambodian postwar media reconstruction and the geopolitics of technology / Margaret Jack.
Description: Cambridge, Massachusetts : The MIT Press, [2023] | Series: Labor and technology | Includes bibliographical references and index.
Identifiers: LCCN 2022034621 (print) | LCCN 2022034622 (ebook) | ISBN 9780262545389 (paperback) | ISBN 9780262374101 (epub) | ISBN 9780262374095 (pdf)
Subjects: LCSH: Postwar reconstruction—Social aspects—Cambodia. | Collective memory in mass media. | Technology—Political aspects—Cambodia. | Digital media—Social aspects—Cambodia. | Digital media—Political aspects—Cambodia.
Classification: LCC JZ5584.C16 J34 2023 (print) | LCC JZ5584.C16 (ebook) | DDC 959.604/3—dc23/eng/20220930
LC record available at https://lccn.loc.gov/2022034621
LC ebook record available at https://lccn.loc.gov/2022034622

10 9 8 7 6 5 4 3 2 1

CONTENTS

Acknowledgments vii
Key Dates in Cambodia 1953–2018 xi

INTRODUCTION: INFRASTRUCTURAL RESTITUTION 1

I MEDIA HISTORY 23

1 INFRASTRUCTURAL HISTORY: AUDIOVISUAL MEDIA IN THE SANGKUM REASTR NIYUM (1955–1970) 25

INTERLUDE: WARTIME MEDIA 49

2 MEDIA CULTURES OF THE PEOPLE'S REPUBLIC OF KAMPUCHEA (1979–1991) 55

INTERLUDE: PEACE TALKS 81

3 THE VIOLENCE OF DEMOCRATIC MEDIA: MEDIA INFRASTRUCTURE TRANSITIONS IN 1990S CAMBODIA 85

II CONTEMPORARY MEDIA RECONSTRUCTION 107

4 MEDIA RUINS 109

5 DISINTEGRATION NOISE: PREAH SORYA'S FILM RECONSTRUCTION AND HEALING THE "WAY OF THE HEART" (*PHLAUV CHETT*) 135

6 MAKING MEMORY LIVE: INTERNET TOOLS OF CAMBODIAN MEDIA HISTORY 159

CONCLUSION 185

Notes 195
Bibliography 225
Index 241

ACKNOWLEDGMENTS

Writing this book was a collective activity. My biggest thanks go to those whose stories are collected here. I am deeply grateful to all of my participants for sharing with me and I hope this book does justice to your work and histories. In particular, I recognize Loak Chayy, Mao Ayeuth, and Khvay Loeung, who have died in the period I spent writing this book. I am so appreciative that I was able to talk to them and learn from them before they passed.

Thank you to Winifred Poster for her guidance on and support for this book and for all she does for the labor and technology reading group. This global community has been an important influence on my intellectual development as a graduate student and beyond. At MIT Press, thank you to Katie Helke, Laura Keeler, and the rest of the production team, who walked me through the steps of being a first-time author and were clear in their editorial suggestions.

I am grateful to my PhD advisor, Steven Jackson, who read many drafts of this manuscript while it was in dissertation form. I also thank my supportive and generous PhD committee: Marina Welker, Karen Levy, Tamara Loos, and Nicola Dell, all of whom brought distinct intellectual contributions to this project.

Thank you to Ingrid Erickson and Melissa Mazmanian for your mentorship and support in 2020–2023. Through our collaboration, I continued to learn new techniques about grounded theory, design, and ethnography as I worked on the final stages of this book.

Thank you to my friends, my collaborators, and the creative people who inspire me in Phnom Penh and Battambang: Lyno Vuth, Meta Moeung, Lyna Kourn, Dara Kong, Davy Chou, Kanitha Tith, Many Sin, Daniel Mattes, Roger Nelson, Rithy Thul, Sela Thul, Kimsru Duth, Thavry Thon, Sopheak

Chann, Socheat Nhip, Pang Sovannaroth, Masy Sou, Melannie Moussard, Channé Suy Lan, Javier Sola, Pen Sereypagna, and Loak Sok. I was lucky to have the space and time with such an inspiring group of people at events, at parties, and in coworking spaces. I have learned a great deal from each of you; I thank you for welcoming me. Thanks in particular to Wynd Veayoo and Hang Jeny for their care.

To my Khmer language teachers Anakkrou Roat, Anakkrou Hannah, Loakrou Socheat, and Anakkrou Channe: *aakun chhareun!* Thank you to those who helped with interpretation, research assistance, and translation: Sokmean Srin, Nehru Ski, Lyno Vuth, Sopheak Chann, Lyna Kourn, Sokanga Hun, Masy Sou, Nhip Socheat, Pang Sovannaroth, Kimsru Duth, and Dara Kong. Thank you to Catriona Miller for her help in romanizing the Khmer words used in this book.

Thank you to my colleagues at CHAI, particularly Emily Welle, Kiira Gustafson, Aude Wilhelm, Joe Novotny, Aman Singh, Mike van der Ven, and Caroline Barrett. They gave me concrete and practical guidance as I was first learning to do qualitative research in rural Cambodia. Thank you to the scholars Alexandra Delfarno, Catriona Miller, Katherine Culver, and Maryann Bylander, who supported me through the personal and professional challenges of doing ethnographic and archival work in Phnom Penh. Thanks to my Phnom Penh–based yoga teacher, Lana Yang, my Vipassana teacher, Beth Goldring, my nonviolent communication teacher, Kathrin Schmidt, and the organizers of "artshram," Steph Mosaik and Raj Bhalla. How lucky I was to learn so much about how to live in the process of writing this book.

Thank you to our 2014–2016 lab at Gates Hall in Ithaca: Vera Khovanskaya, Lara Houston, Stephanie Steinhardt, Samir Passi, Leo Kang, and Ishtiaque Ahmed. What a generative moment of intellectual community! I learned about the intricacies of infrastructure through our conversations. Thank you to Helen Nissenbaum, Jessie Taft, Michael Byrne, and the groups of postdocs and doctoral students between 2019 and 2022 at the Digital Life Initiative. Our discussions helped me come to understand the nuanced stakes in questions of privacy and ethics of emerging technology.

I am grateful for my writing partners and coworkers in New York between 2020 and 2022, particularly Meg Young, Rosie Bellini, Charanya R., and Cindy Lin. I thank my virtual coworkers, too, including Seyram Avle, Sharath Chandra Ram, Catalina Alzate, Leah Horgan, M. C. Forrelle, and Anupriya Tuli. During periods of pandemic isolation, our working sessions kept me creatively and intellectually engaged while also increasing my morale.

Thank you to the 45th composters in Sunnyside, particularly Austin Kim, Lauren Berke, Kristina Baines, Victoria Costa, Radoslaw Wojciechowski, Kaylyn Kilkuski, Bob Braun, and Alexa Lehoczki. Our conversations about art, film, and books over magical dirt helped me through difficult times during this project.

I am also grateful for my funders. I completed this research with the financial support of the US National Science Foundation, a US Foreign Language and Area Studies Fellowship, a Women in Technology New York Fellowship, and a Digital Life Initiative Fellowship.

Thank you to my family: Mom, Dad, Annie, Sam, Mike, and Nina. A special thanks to Louis, Lana, Bill, and Marie! Your arrivals and growth in the past seven years have brought me so much joy. Thank you to Cyrus Mossavar-Rahmani.

Thank you finally to Ingrid Muan, whose papers and writings are so thoughtful and have been so valuable to this project.

KEY DATES IN CAMBODIA 1953–2018

November 9, 1953: Norodom Sihanouk declares independence from French Indochina

1955–1970: Sangkum Reastr Niyum period—Norodom Sihanouk head of state

1969–1970: America bombs Cambodia as part of the American war in Vietnam

March 18, 1970: Sihanouk is deposed by right-wing general Lon Nol

1970–1975: Civil war between Lon Nol–led Khmer Republic and communist Khmer Rouge

April 17, 1975–January 7, 1979: Khmer Rouge regime rules Phnom Penh, 1.7–2.5 million die from execution, starvation, and disease

1979–1992: Communist People's Republic of Kampuchea, with Vietnamese occupation until 1989

February 1992–September 1993: United Nations Transitional Authority Cambodia; privatization period and opening to the West begins

July 5–6, 1997: Co-premier Hun Sen of the Cambodian People's Party ousts Norodom Ranariddh of FUNCINPEC. Fighting continues in the provinces until September.

April 15, 1998: Pol Pot dies

November 16, 2017: Cambodian Supreme Court dissolves Cambodian National Rescue Party

INTRODUCTION: INFRASTRUCTURAL RESTITUTION

On a sunny and hot Phnom Penh Saturday morning in August 2017, I parked my motorbike at the Royal University of Fine Arts (RUFA) to join a heritage cinema tour. The organizers were the youth group Roung Kon ("cinema"), and our tour guide introduced herself as Kagna, a recent architecture graduate. She planned for us to visit twelve of the most famous pre-1975 cinemas; none of these was still running as a cinema and all rested in various states of demolition. We started our walk at the nearby Cinéstar, the remains of an early twentieth-century cinema. The roofless shell of a building held a line of colorful clothes, a brush and bucket for washing, an unused food cart, a plastic chair, a few bicycles, and a very large palm tree growing out of the concrete floor. Above a wall-less room (what was once a functional washroom) in the far corner of the building, a three-meter by four-meter block of white was painted onto the old cement. Here a graffiti artist had rendered a blue-tone portrait of a young girl. Her eyes look up, her mouth is slightly open, and her hair, parted to the side with a swept-away bang, shadows the left side of her face. She looks surprised or afraid. She might even be watching a scary moment in a film.

For the members of Roung Kon, revisiting these sites was critical for their generation; they were important social and cultural spaces during the "golden age," the extraordinary time of the arts in Cambodia after independence from France (1953) but before the worst of the civil war (1970–1975) and Khmer Rouge regime (1975–1979). "Neighborhoods are even known by old cinemas," Kagna explained. These cinemas were symbols of meaningful memories of entertainment and the arts. Another member of the group, Daro, later told me that he imagined older people walking past these spaces and thinking about the happy times they had there. "I don't have the memories [myself], but I can feel them." The mystery around the cinemas drew Roung Kon in. "Most of the cinemas are gone but we can see some structures that still exist. We want to know more about them," Daro said.

Once these cinemas represented refuge: spaces to build social, in-person networks of trust, entertainment, and joy, critical in the tumultuous political history of Cambodia. These kinds of spaces were needed again in the years of 2017 and 2018. In a rapidly urbanizing Phnom Penh, heritage buildings like these were being torn down at a rapid pace. Casinos were being built in lieu of public, state-sponsored cultural or green space. Cambodia had also entered a politically sensitive period preceding the Cambodian general election (July 29, 2018), which human rights advocates have widely criticized for representing a pivot toward authoritarianism after twenty-five years of international support for democratization efforts. In November 2017, the Cambodian Supreme Court dissolved the primary opposition party, the Cambodian National Rescue Party. These political events were coupled with increased regulation of media and arrests for oppositional speech.

Media have long been linked to political power in Cambodia. Cinéstar was one of the first cinemas in Phnom Penh. Built within 100 meters of the Royal Palace and the National Museum, Cinéstar was once a royal theater, playing almost exclusively French films during the colonial period and early postcolonial period. Norodom Sihanouk was interested in film from an early age growing up in the palace with arts-loving parents. Sihanouk then ruled during the so-called Cambodian golden age of arts, the Sangkum Reastr Niyum (1955–1970) period, when Sihanouk himself became a filmmaker.

Though a time of flourishing arts, all was not well in Cambodia during the Sangkum Reastr Niyum period; instead, it was a time of intense geopolitical conflict and insupportable levels of domestic inequality. Amid increasing communist organizing, a US-backed coup threw Sihanouk out of power in 1970. The country was then embroiled in civil war between the Khmer Republic and the communist Khmer Rouge. The Khmer Rouge took power in April 1975 and ruled until January 1979. During this period, approximately a quarter of the Cambodian population died of execution, starvation, and disease.

This book's central argument is that Cambodian media creators and technologists perform *infrastructural restitution*, the creative reconstruction of historical media artifacts and infrastructures, as a way to access positive affect about the national past and work toward political action. This work is particularly meaningful in Cambodia because the Khmer Rouge regime systematically destroyed cultural artifacts, including media (radio, film, photography, and music) from before their rule. In the decades of conflict following the Khmer Rouge, much of the remaining historical material decayed or was lost. Restoring these artifacts is significant to young media creators because they capture a time in Cambodia of peace, cultural flourishing, and beauty. Working with historical material also provides an avenue for remembrance of positive cultural histories and imagining building a better Cambodia, particularly relevant in an increasingly authoritarian and information-controlled environment.

Much of the contemporary work of infrastructural restitution is done on digital tools, which have rapidly grown in popularity in the past decade. In 2013, the United Nations International Telecommunications Union estimated that 7 percent of Cambodians had access to an internet connection. By 2019, a startup ecosystem report reported that 84 percent of Cambodians had access to the internet.[1] The majority of internet use in the country is the use of Facebook on internet-enabled phones.[2] Top tech analysts in Cambodia in 2017–2018 reported that Facebook was by far the most actively used web platform in the country. Approximately every internet user in Cambodia has a Facebook account and many confuse the internet for Facebook, a semi-monopoly that exists in many countries, particularly in the Global South.[3] This dramatic technological transition has had a profound impact on the Cambodian economy, culture, and politics.[4]

I began working in Phnom Penh, Cambodia, in January 2014 as a market analyst for a nongovernmental organization (NGO), researching the ways that digital technologies were being used in health care settings in rural parts of the country. I witnessed the beginning of this dramatic growth of the internet across the country. When I first arrived, fresh from working as a financial analyst in Silicon Valley, I started looking for "innovation" in stereotypical technology centers like coworking spaces and startup weekends based in the capital. As I continued living in Phnom Penh and became embedded in the city's vibrant cultural institutions as a researcher in the following years, I realized that the most inspiring technology work I was seeing was coming out of places I wasn't expecting—community art spaces and archives—and in forms that were emerging from Cambodia's particular histories of art, as well as its legacies of violence. These observations led to the central question of this book: What is the role of media in post-conflict reconstruction?

Healing from past violence and using history to interpret the contemporary political world are important themes all over the world in 2021. In the United States, we are in the midst of pressing social dialogue about the ways that historical violence continues to be displayed through material culture (e.g., Confederate statues). We are collectively rethinking the ways that we tell history and how we deploy history to call for social change. These sorts of cultural reckonings in post-conflict environments have occurred in other parts of the world, for example, Latin America, South Africa, or Eastern Europe. This book contributes to these conversations and considers the particular role of digital media in processing the national past and moving forward.

Cambodia provides a compelling setting to explore the central themes of this book since its particular histories of conflict impact the use of digital tools in clear and multiple ways. When I refer to the "conflict" in Cambodia, I refer to the entire Cambodian war period, lasting from the late 1960s when the United States bombed Cambodia as a part of its war in Vietnam through Pol Pot's death in 1998.[5] Divisions between warring factions within Cambodia were deeply entrenched and continue to be, to some extent, unresolved. The line between victim and perpetrator was not always clear, especially because the Khmer Rouge recruited many child soldiers and the conditions of postcoloniality, poverty, war, and violence in the country before the regime made the ethics of and justifications for joining the Khmer Rouge complex. Though the Khmer Rouge atrocities happened approximately forty years ago, the ensuing conflict has delayed recovery and justice, including institutional justice. Members of the Khmer Rouge kept power long after the end of the regime, or were restored to power during reconciliations in the 1990s. Today victims and perpetrators often live in the same communities.[6]

Though the entire war period deeply affected Cambodia, the Khmer Rouge era was acutely disruptive and destructive to the country's art and culture. The regime, in calling for a radical restructuring of society and a return to Year Zero, attempted to destroy all cultural remembrance of earlier times in Cambodia. They specifically targeted intellectual and cultural figures for execution and tried to destroy the art, media, and libraries created before the regime. Former residents of Phnom Penh who were sent into forced labor camps in the rural districts were not allowed to sing or play songs from the old regime, though many said that singing old songs in the fields remained a common form of resistance. These emotionally violent censorship policies continued through part of the People's Republic of Kampuchea (PRK) period following the Khmer Rouge.[7]

The core interest of this book is how practices of memory and healing in Cambodia occur on and through media, the communication outlets or tools used to store and deliver information and data. *Infrastructural restitution*—the creative reconstruction of media infrastructures—plays an important role in assuaging traumas and building new futures in Cambodia, a commemorative and healing process not yet fully accounted for in scholarship on Cambodian experiences or on collective trauma more broadly. This concept allows us to query the special relationship between work, memory, and materiality in a postcolonial and post-conflict setting and understand creative reconstruction as a form of healing and political action. Infrastructural restitution is a largely unrecognized and deeply affective form of labor that bridges social and technical, material and ephemeral, and online and offline worlds.

Restitution has two common, lay definitions: (1) the restoration of something to its original state, including things that are lost or stolen, and (2) the act of recompensing for injury or loss. Restitution takes on an air of justice; it is often understood as a physical adjustment to fix something historically unjust. In US law, the law of restitution refers to a form of compensation that can be awarded in some cases if there's a breach of contract. The purpose is to prevent one party from being unjustly enriched at the expense of another. Rather than to punish, the goal of the law is only to put the wronged person back into the position they were in before the breach happened. We can see this sense of restitution in action in the rightful return of looted Cambodian cultural heritage materials from art institutions in the West.[8]

I use the term *restitution* to describe a set of actions used to return to, reflect on, or honor a pre-conflict state. As in US law, the goal of restitution is not to punish but to go back to a previous state before the conflict. As opposed to in law, I am referring not to court-imposed compensation but rather to survivors or their ancestors working for a return to a previous condition. In defining restitution in this agentic way, I am inspired by Jihan El-Tahri's art project "Complexifying Restitution," which brings together artists and filmmakers to reappropriate archives "as a tool to reinterpret our current reality and imagine an alternative future."[9] El-Tahri's definition of restitution empowers contemporary readers of history in reaction to the way that colonized subjects had a limited degree of self-representation.

The politics of restitution can be fraught and contested as restitution encompasses both conservative and progressive impulses. Returning to a pre-conflict state often challenges the contemporary status quo and can be a form of progressive social action. Yet restitution can also harken back to and occasionally romanticize historical inequities and violence or entrench old power structures. The political meanings of restitution differ according to perspective and change through time, and we will see a variety of political meanings emerging from the work of infrastructural restitution.[10]

The kinds of restitution this book describes are specifically *infrastructural*. Building on scholarship of infrastructure, its framing addresses the interdependence of materiality, work, and structural power within the work of restitution. These include forms of coordinated action between people and machines, transnational sociotechnical systems, and the power dynamics embedded in these assemblages. This definition emerges from the sociology and anthropology of infrastructure, and my theoretical contributions are most directly related to conversations in those fields and related fields like science and technology studies, information communication technology for development, and human-computer interaction.

Scholars of infrastructure have long paid attention to the relationship between form and content, and the myriad impacts of infrastructural materiality. For example, critical media infrastructure scholars acknowledge the material hardware that make up media and information infrastructures, and the impacts of this materiality on information content and delivery.[11] These scholars break myths of the ephemerality of information including the rhetorical framing of so-called cloud data storage.[12] They point to the intense material hardware required for digital information and the energetic and environmental consequences of this information.[13] Drawing on this literature, throughout the book, I pay attention to the *stuff* of infrastructure—transistors, film reels, air conditioners, cinemas, smartphones—and the way that this stuff takes on important political, affective, and symbolic meaning.

Conditions of techno-materiality act as limitations and inspiration for infrastructural restitution. Working from critical media infrastructure scholars like Parks and Starosielski, and Larkin, I pay attention to the physical qualities of media technologies and how they move through space.[14] This infrastructural approach gives a fresh understanding of the geopolitics of technology, highlighting how materials act as affordances and agents in foreign and domestic politics, moments of cultural encounter, and modes of power. Looking at a history of devices allows us to develop a sense of media use and access that is impossible to develop through content analysis alone, particularly in this context where the archival record is patchy or weak. Paying attention to the material qualities of media infrastructures allows us to glean new insight into how media were taken up and used by average media consumers. Taking this approach also gives us a deeper sense of the constraints and possibilities for media artists and the conditions under which they were able to create new kinds of films and radio content. In later chapters, aging media materiality acts as a vector for accessing positive affect about the past.

In contrast to a primarily physical conceptualization of infrastructure, I emphasize that the *work* and the materiality of infrastructure are inexorably tied, an insight that has been developed particularly strongly in the sociological tradition of infrastructure. Leigh Star throughout her career paid attention to the often unrecognized and undervalued work of infrastructure, particularly the kinds of "work that makes work happen."[15] In 1996, Star and Ruhleder theorized that infrastructure is social and technical, global and local, labor and material.[16] This approach underscores that infrastructure does not need to be large or even "technical," though Star does hold onto the material component of infrastructure.[17] Since Star, scholars have highlighted the active, iterative making and remaking of infrastructures in science and technology studies and critical technical scholarship through concepts like "infrastructuring" or "creative infrastructural action."[18] Nguyen's concept of "infrastructural action," for example, describes members of the Vietnamese diaspora bringing cheaper and higher quality smartphones from the United States to Vietnam for family members; this case demonstrates well the ways that infrastructural work can take on dimensions of global scale.[19]

There is a social justice ethic to paying attention to invisible infrastructural work. The concept helps to recognize people whose labor and creative contributions are often overlooked, particularly in technical fields or the innovation economy. Star points out that the work of librarians, archivists, technicians, nurses, and homemakers can be considered infrastructural in the sense that they provide support for the accomplishment of interrelated tasks.[20] This insight is aligned with Star's complementary work on "residual categories."[21] This concept captures the things or people who are left out of bureaucratic forms because they exist outside or between categories, and the psychological and material outcomes of that marginalization.

Invisible infrastructural work often takes on a character of care. For instance, Star and Strauss suggest that nuns saying prayers for the sick and troubled practitioners can be understood as a kind of infrastructural work: this contemplative work helps the community to function.[22] Self-care, too, is a kind of work when it allows us to do other work. In one extreme example, Star and Strauss borrow a story from Morrison's *Beloved* of the slave Sixo who steals a pig from his master. He justifies that this theft is a form of work, since feeding himself will allow him to produce more in the fields. Labor scholars have long recognized invisible forms of affective labor, or work that is done to produce emotional experiences for others.[23] We can see this kind of work done, for example, by customer service agents who pretend to be a different race or nationality in order to produce comfort for people on the other end of a line.[24] Lindtner describes the role of "happiness workers" who are responsible for lifting the affective mood of a coworking space so that independent workers can more effectively do the work they need to do.[25]

Infrastructural restitution is fundamentally a form of work (restitution is an action), but a highly affective and often invisible form of work. Often infrastructural restitution takes on forms that are previously recognizable as infrastructural (e.g., sophisticated transnational logistical work). I also expand the traditional concept of infrastructural work to include processes of emotional healing through creative and cooperative action. These forms of work are infrastructural work because they build a more emotionally healed baseline from which media creators and technology producers can move forward and dream toward new futures. It is part of the healing work that allows *any* work to happen. Recognizing the affective and, at times, even metaphysical level of infrastructure helps us see this highly affective form of work as infrastructural.

Infrastructure studies also point to the *relational* character of infrastructure. That is, as much as we consider the things and work of infrastructure, we must also focus on the relationships inherent to infrastructure and the ways that people differentially interact with infrastructure. Infrastructure is always embedded in social relations; it can never be separate from local hierarchy and global politics.[26] These already-established power dynamics are lived in and through infrastructure. For example, in "The Ethnography of an Infrastructure," Star points to city water infrastructure as a set of relationships instead of a set of things. The infrastructure has a variety of meanings depending on one's relationship to the infrastructure. She explains, "The cook considers the water system as working infrastructure integral to making dinner, but for the city planner or the plumber, [the water system] is a variable in a complex planning process or a target for repair."[27] In this way, the infrastructural lens gives us an avenue into analyzing violence, power, and structural inequality. Inside an infrastructure, one person's affordance is another person's barrier.[28] Infrastructure theory has thus given a potent lens to analyze critically questions of circulating technology.

Infrastructural restitution is a relational practice that occurs within power structures. Conditions of privilege and power have led to infrastructural violence through media technologies, under all kinds of political arrangements. Today the postcolonial politics of the platforms (Facebook and the smartphone app) on which much of the infrastructural restitution is done marks a continuation of some colonial politics and the impulses of the "imperial archive" and bound the types of work that many artists do.[29] As I will show, independent media creators have long co-opted imperialist and nationalist media infrastructures for their own purposes. Contemporary media creators, too, creatively appropriate transnational technology platforms that exist only within an unequal matrix of transnational and capitalist power dynamics. Actions of infrastructural restitution occur within structures of imperialism and global capitalism while often promoting justice agendas. These tensions and contradictions abound within infrastructural restitution, and this work takes on complex matrices of affect.

In order to build the concept of infrastructural restitution, I bridge insights from infrastructure scholarship with literature at the intersection of media and memory throughout the chapters of this book. Bringing these bodies of scholarship together helps us to see that infrastructure is comprised of and helps to constitute immaterial things like affective and psychological experience, including memory and emotions, and extra-human immaterial things like ghosts and hauntings. Media infrastructure takes on a special character in Cambodia, where the memories of colonialism, authoritarian politics, and conflict are painfully significant and sometimes suffocating in their unrelenting presence. In many postcolonial settings, memory often remains unsettled and memory conflicts play out in material things (such as in the urban form of the city) through the contestation over heritage, identity, and difference.[30] Avery Gordon writes that the past—particularly those legacies of colonialism and slavery that she calls *ghostly matters* or *hauntings*—is constitutive of the present and must be attended to.[31]

In Cambodia, Edwards argues that it was the intersection of French and indigenous worldviews in the 1860s that first fostered national identity.[32] "Cambodian" culture is an ever-changing thing; however, this style of nationalism fundamentally shaped the ideologies of political leaders post-independence, from Sihanouk to Lon Nol to Pol Pot to Hun Sen, including its emphasis on replicating imagery of Angkor Wat. Each of these leaders adopted a myth that Cambodians are changeless and propagated a backward-looking, Angkor-centric nationalism. They did so by cordoning off the colonial era as somehow inauthentic.

During the Khmer Rouge period, the past was (nominally) banned—nostalgia was renamed "memory sickness."[33] Pre–Khmer Rouge songs, books, culture, and arts were forbidden, as was money. This memory sickness was part of the Khmer Rouge policy of *kamtech*, "to destroy and then to erase all trace: to reduce to dust," or eliminating all forms of individualism to preserve the primacy of Angkar (or "the organization," the name that the Khmer Rouge leadership used to refer to itself).[34] But Pol Pot and Khmer Rouge leadership actually placed themselves within the *longue durée* of Cambodian history and compared themselves to the "Original Khmer" (*Khmer daem*), a figure who predated the French arrival in Cambodia. They therefore did not abolish history but instead erased colonialism from Cambodian history, continuing to replicate and rely on the emblem of Angkor Wat.[35] These contests around "true Khmer-ness" play a role in the conservative nature of infrastructural restitution, as some media workers lean on past media artifacts to make arguments about political legitimacy and nationalism.

Post-conflict memory is also central to this story, and in Cambodia, it is a complex emotional experience that is at the same time collective and deeply individual. Collective trauma like terrorist attacks, war, and genocide can powerfully disturb people's worldviews and challenge the common basic belief that the world is benevolent, predictable, and meaningful.[36] In post-conflict settings such as Guatemala, Germany, Vietnam, and South Africa, memory has become a particularly subjective, political, and contested terrain. Scholars have described how plural memories of the past become points of cultural tension, often played out through material things like monuments.[37] Schwenkel's concept of "recombinant history" describes the disputed collective understandings of history in the Southeast Asian context.[38] Different memories are "not additive but dialectical" so their recombination requires fitting together competing representations.[39] Cambodian memory work happens in many forms, for instance, through arts, craft, body work, and spiritual practices, and this book is in conversation with other scholars of commemoration in Cambodia such as Kristina Uk, Cathy Schlund-Vials, Khatharya Um, Boreth Ly, Lina Chhun, Ashley Thompson, and Anne Guillou, among others.

The core interest of this book is how practices of memory in Cambodia occur on and through media, the communication outlets or tools used to store and deliver information and data. Through the chapters, I build on scholarship that queries the relationship between media and memory in post-conflict settings like Cambodia. This scholarship addresses ghosts and hauntings as remnants of historical violence in the built world, the power of media to curate collective memory, and the clarifying role of noise in understanding the possibilities of media for emotional healing.

I am also in conversation with scholars in the fields of postcolonial science and technology studies and human-computer interaction who describe and theorize the movement of new digital tools into postcolonial settings. Chan's *Networking Peripheries* shows how digital alternatives flourish in spaces far from centers of global technology such as Silicon Valley. She argues that computing and innovation cultures are rendered locally specific due to language, religious practice, gender dynamics, and other social factors. In *The Charisma Machine*, an ethnography of the One Laptop per Child program, Ames argues that American technology interventions often fail to understand fully or design for conditions and infrastructures outside the United States. She also illuminates problematic power dynamics between MIT-based techno-elites and the global users for whom they purportedly design. Irani's *Chasing Innovation* charts the ways that the racist, nationalist, and class prejudices of the Silicon Valley design sector exacerbate global inequality. In *Prototype Nation*, Lindtner describes the ways that growing distrust in the Western model of development and digital governance led to a change in China's global image from a copycat tech manufacturer to a new frontier of innovation through democratized making.

   Media Ruins builds on these critiques while also offering a story about the ways that technology users make use of tools that are unexpected to their original Silicon Valley designers and in line with their own local needs and environment. *Media Ruins* holds in tension the political economy of transnational platforms and the ways global media creators appropriate these platforms and use them as channels for healing and political action. *Media Ruins* focuses on the ways that history matters in understanding the contemporary geopolitics of a locally specific internet and innovation culture. Cambodia represents an excellent case for interrogating these questions since its histories of trauma and conflict so clearly impact contemporary digital work.

   In order to build these arguments, I offer reflection on the role of media infrastructure in the historical and contemporary geopolitical climate of Cambodia, critical to understanding both the conflict in Cambodia and the dynamics of healing from it. Infrastructures and all of the bits, workers, warehouses, and wires of technology are now some of the key pieces of international relations. Mainstream media often paint the contemporary world as a polarized one with two major models of digital governance. The first is best exemplified by the current Chinese government, which enacts communitarianism, nationalization of platforms, and strict state information control and censorship. The second model of digital governance is most visibly represented by American technocapitalism, where monopolized technology companies obscurely keep data as proprietary for advertising. The extent to which these models are truly different is questionable: Chinese companies exploit workers and use data for information capitalism, and corporate monopolies in the United States are deeply entwined in state structures. Though emerging from supposedly different ideological places, the reality of these two governance models often appears quite similar, with both enacting forms of surveillance capitalism.[40] Both countries are competing for global users on their domestic internet platforms as a form of capitalist expansion.

   Cambodia is one of the liminal media points that is emerging at the margins of these competing epistemologies about global digitization amid empire shift. During the run-up to the 2018 national election, Chinese influence grew in matters of digital media policy, while Cambodian relations with the United States and European Union chilled. Both regions continue to have major corporate spheres of influence in Cambodia and the Southeast Asian region at large. Chinese corporate interests are dominant in new telecom infrastructures such as 5G, but the American corporate platform Facebook remains by far the most popular way for most Cambodians to use the internet.

The Cambodian government, in line with regional trends and based on the Chinese model, is moving toward technological "localization," and is trying to rein in American influence in technology and regain control over the "digital economy" and data flows. This move toward control and sovereignty, though sometimes couched in the language of decolonization, is also tied to a pivot toward authoritarian politics. Cambodian independent media creators are working again within a context of authoritarian information control similar to the environment artists lived through in the early Cold War.

The year 2017, when I was conducting the majority of the fieldwork for this book, marked a resurgence of censorship and control policies in Cambodia. The government began to use digital platforms as governing tools, enacting new internet-enabled surveillance technologies and strategies. That August, the Cambodian government dramatically curtailed freedom of speech on the internet and in the media sector more broadly by systematically closing major critical media outlets, including newspapers and radio stations. The government also established new telecommunication laws and began making arrests based on critical Facebook activity. This closure in the media sector was coupled with the dissolution of the primary opposition party, the arrest of its president, and the fleeing of many opposition leaders.

This book shows the ways that digital media technologies have broadly become tools for global geopolitical meddling and authoritarianism in Cambodia, and that this trend is part of a clear historical trajectory. I describe the historical legacies of four important contemporary media governance trends: foreign interference (United States Information Service [USIS], 1955–1963), media authoritarianism (starting from the Sangkum Reastr Niyum period, 1955–1970, particularly 1963–1970), South-South geopolitics (including the Vietnamese occupation during the 1980s), and Westernized neoliberal development (starting from the United Nations Transitional

Authority of Cambodia, 1992–1993). In sum, the book shows that media have long been tightly linked to political power in Cambodia.

In addition to detailing the national-level geopolitics of technology we see in Cambodia and their historical roots, this book also shows how Cambodian young media creators respond to these issues with infrastructural restitution as a form of political action. I show the ways that Cambodian media creators use the praise of historical moments as a way to call subtly for change. I also demonstrate how Cambodian media creators have established or preserved spaces of communication and independent thought. Brick and mortar spaces of trust have become only more important as the internet becomes widely understood as an unsafe place.

Politically, this work is motivated by the feminist imperative to describe what kinds of care practices work amid broken systems of authoritarianism, imperial legacies, racial hatred, violence, and techno-dystopianism. It is resonant with the feminist attention to "glitch politics": studying and describing how people (particularly those in progressive grassroots movements) "thrive otherwise" against the oppressive and exploitative aspects of our digitally mediated worlds.[41] This feminist approach is aligned with Tsing's narrative of the hardy cultures of Southeast Asian diaspora mushroom hunters in the capitalist ruins of the Pacific Northwest or Escobar's attention to the interdependent, autonomous communities of the Cauca River Valley of Southwest Colombia who are designing a transition to a radically plural way of being.[42] Dwelling on the past might lead us to be stuck in an overwhelming sense of sadness and tragedy. Instead, we can choose to look at the ways that media have helped new creators process feelings and politically engage with new ways of thriving. This study of infrastructural restitution attends both to the ways that media becomes a tool for nationalist, state power and to personal, affective, and future-building work.

Media Ruins works between the modes of history and ethnography, tracing continuities between older forms of media (film and radio) and contemporary commemorative media projects, which use digital tools in conjunction with older media forms. Instead of seeing the digital as a break or gap, this book insists on historicizing the digital to see its social impacts as part of longer-term processes of commemoration, governance, and cycles of violence. I illuminate the relationships between colonization, technological imaginations, and nationalism and make linkages between contemporary media infrastructures and their historical precedents.[43] I therefore move between three different methodological voices—archival-based history, oral history, and ethnographic description—in order to shed light on its narrative arc from various perspectives and give its story multiple dimensions.

This book is based on six years of historical research and ethnographic participation in the arts and technology communities of Phnom Penh and Battambang, Cambodia, including twenty months of research from June 2017 until January 2019. I conducted over 100 audio-recorded interviews (in Khmer and English) with technology producers, media artists, and active social media users across rural-urban, age, and income demographics in Cambodia, and had countless informal conversations around issues of historical and emerging media technologies. I also participated in and helped organize participatory art and design events, and attended approximately two dozen events on historical memory, art, and technology. The cases in this book are selected from this collection of experiences and represent ethnographic relationships built over a six-year period. The historical sections of the book are based on document review in six archives in Cambodia and in the United States. I supplemented my archival research with oral histories with Cambodian elders and a review of secondary historical sources.

I consider theory and method together, and so my choice to use an interpretive, critical, and historical approach to analyze questions of contemporary emerging technology is itself significant. It says something about how I understand the interdependence of cultural, political, social, and historical worlds and computing ones. Qualitative research, particularly ethnography, gives us an opportunity to make useful stories about technical worlds and helps us make sense of the ways the "technical" impacts and is impacted by the "social."

Ethnography is particularly well suited for trying to get at the messiness, incoherence, and instability in things that I am interested in (like infrastructures), which tend to be seen from the outside as monolithic things.[44] I see the "field" as a metaphor to refer to the places, sites, and times that helped me get answers to my research questions.[45] My research questions, however, couldn't be answered by sitting in a particular location over the course of my research period. Many of the phenomena I wanted to understand were spread out, part-time projects, done during my participants' nights and weekends at home, at coffeeshops, and at coworking spaces, online and offline. I therefore needed to construct my research site as a "network."[46] This strategy was particularly suited to helping me study the interrelationships between online and offline spaces, work, and social life. I bought a motorbike in June 2017 and learned how to ride it so I could get around Phnom Penh more easily and go where I needed to for research. I regularly attended the abundance of arts and technology events from art history talks to tech startup weekends and social enterprise networking nights, which were mostly advertised on Facebook and held on weekends and evenings. These events exposed me to new kinds of information about technology and the arts of memory in Cambodia, and also gave me other spaces in which to interact with and build relationships with a community of artists, technologists, and academics interested in my research questions.

Facebook was an invaluable tool in my research process. Throughout the course of the project, Facebook became an object of study, a tool of research, and a site for participant observation. Almost all of my younger participants actively used Facebook, and I used Facebook to set up meetings using Messenger, to call participants with follow-up questions, and to communicate in other various ways (e.g., posting on others' pages or liking their photos). I used Facebook to better understand social relations between various social groups and subgroups and to keep up with collaborators without visiting them in person. I was able to find primary sources (e.g., some films) better on Facebook than in institutional archives. Facebook lists nearly all public events in Phnom Penh and helped me stay involved in many communities. As Burrell suggests, multi-sited ethnography now necessarily includes online spaces in various ways since online life is inseparable from offline life.[47] Since returning to the United States, Facebook has been a way to observe Phnom Penh–based happenings and keep in contact with friends and colleagues in Cambodia. It has both enlarged the sites of research and made them more tightly linked to each other.

Chronicling digital innovation in Cambodia, a postcolonial, post-conflict setting, adds an important counterpoint to many American histories of technology, which often simplistically and monochromatically relate moments of Western techno-scientific innovation in elite academic settings or technology centers like Silicon Valley. Though the so-called techlash—including critical takes on the bias inherent to algorithmic systems or the privacy concerns with the big tech business models—has started to seep into the American and European press and consciousness, there are far fewer conversations about the impacts of tech on daily life, security, and social infrastructures in other parts of the world. When these are presented, they often are packaged in patronizing or sensationalized ways, in extreme settings.[48] This book explores the monopolistic and exploitative nature of Silicon Valley–generated tools, but without stripping users of their agency and creative capacity. This book also puts these new transnational tools into a historical perspective, and shows that the US government media initiatives in the early Cold War period are resonant with contemporary expansionist corporate policies.

My positionality as a white, American, and middle-class researcher in Cambodia has important limits. White privilege exists in Phnom Penh in many clear and some subtle ways. Wealth and power are now associated with the infrastructure of the international development organizations that populated the city during and after the United Nations Transitional Authority of Cambodia. My best work has come only through dialogue with Cambodian friends and researchers at all stages of research, from problem delineation to execution and analysis. I do not claim to know the experience of Cambodia in the way that my Cambodian or Cambodian diasporic participants do. This book is a product of my deep listening to my participants, interpreted through my own embodied, historical, and psychological experience. I also feel some responsibility, as an American and as a former tech worker, to research and write about the problematic kinds of work that American corporate and government entities have done in settings around the world.

I did in-depth and audio-recorded interviews of major actors in my case studies in homes, coffee shops and offices, often with multiple visits or conversations. These interviews were semi-structured. I wrote the interview protocol specifically for each person, depending on my specific interest in their experience. I also did strategic interviewing and oral histories among older Cambodians (forty-plus) who had memories of 1980s and 1990s Cambodia and were involved in reconstruction efforts and/or had television and radios during the PRK period. I also interviewed people in the contemporary technology startup community and in the radio and film industries to get a sense of major issues and concerns in these sectors.

I sometimes brought a research assistant with me for Khmer-language interviews, though many interviews I conducted alone in English. Lyna Kourn, Sokanga Hun, Masy Sou, Nhip Socheat, Pang Sovannaroth, Kimsru Duth, and Dara Kong all acted as research assistants/interpreters for interviews at various points during the research period.⁴⁹ Interviews that included three people were often easier and more effective on a number of levels than interviews with two people; we could develop a rapport and familiarity more easily and I often felt more comfortable having a friend with me. Interviewing with a Cambodian interpreter also sometimes helped the interview subject feel more comfortable because of the cultural familiarity of the Cambodian interviewer. I also appreciated having a research assistant to process interviews with afterward if there was particularly sensitive or important information shared.

I used English, French, and Khmer language for data collection. Many arts and technology events are in English, the dominant international language of Phnom Penh, and many of my participants speak English. I read archival documents from the Cold War period in French, Sihanouk's primary language of communication. I took two years of formal Khmer language at Cornell and an intensive course in the summer of 2017 in Phnom Penh. I also had a Phnom Penh–based tutor named Y Socheat, whom I met with three times a week for an hour for language learning in 2014, 2016, and 2017–2018. I continued learning new vocabulary words and we spoke together to increase my fluency in understanding and speaking. Socheat also helped me with translations of textual documents or audiovisual Khmer-language materials I found in the archives. This project is substantially based on original English translations of Khmer and French-language primary sources. I chose to primarily present words in this text as English translations of interviews and textual materials, partially because of challenges in the standardization of Khmer romanization.⁵⁰

I recruited interview participants through snowball sampling, directed emails, and phone calls, or by showing up at someone's office or workplace and asking for an interview. As with many ethnographies, access posed a challenge at times. Access was easier to younger, more cosmopolitan participants than it was to older, establishment figures, who sometimes required special permission or emails. These older, establishment actors required more intense scheduling and often didn't allow me to "hang out" the way I could with younger people.

I wrote frequently during my research period. In the "Manifesto for Field Work," Willis and Trondman say that half of ethnography is "richly writing up the encounter, respecting, recording, representing at least partly in its own terms, the irreducibility of human experience."⁵¹ I iteratively took several kinds of notes and more formal writing throughout my research periods.[52] I took morning notes in a "personal" notebook—my habit was to write two pages a day of handwritten observations, personal reflections, and notes on my own mood and well-being. I kept a second notebook with me through the day for more "professional" jottings, notes while in meetings, analytical thoughts, or descriptive observations during the day. I tried as much as I could to transcribe the second ("professional") notebook into a computer in the evening, particularly after a fruitful research day, and added more analytical thoughts, to form analytical memos based on theory and research questions in the grounded theory tradition.[53] I wrote analytical reflections during downtime when I wasn't at events or while hanging out at coworking spaces. Part of my impetus for writing so much was that I wanted to make sense of what I was learning and to make sure I was continuing to collect the most important information for my writing. Sometimes writing felt like a break from the social parts of research, which could be exhausting for the introverted parts of myself.

The book is broken into two distinct halves. Part I is historical, describing the film and radio cultures of Cambodia from 1955 through the 1990s and demonstrating the tight linkages between media and politics in Cambodia through regime changes. The chapters in part I describe the construction of the media sector in Cambodia and its reconstruction after the violence of the Khmer Rouge. Part II describes contemporary media commemoration projects and is based on my ethnographic participation in the arts and technology communities of Phnom Penh. This part describes contemporary infrastructural restitution, and each chapter highlights a different dimension of the concept.

Chapter 1 provides critical context on the origins and development of film and radio infrastructures in Cambodia before the destruction of the Khmer Rouge period, and the moment of (relative) peace and cultural flourishing that contemporary media reconstructors reference. Norodom Sihanouk declared Cambodia independent from French Indochina in November 1953 and then established effective one-party rule under his Sangkum Reastr Niyum party from 1955 until 1970, during which time Cambodia was officially neutral in matters of Cold War policy. This was also a period of thriving Khmer modern art, architecture, music, and film—and deep inequality.[54] The chapter argues that the investment of the United States Information Service (USIS) in film and radio in Cambodia between 1955 and 1963 strengthened and supported national audiovisual media infrastructures. Sihanouk then deployed these same infrastructures to undergird his authoritarian power from 1963 to 1970.

The second chapter introduces the concept of infrastructural restitution empirically and describes the work, materiality, and relationality of media reconstruction during the PRK period (1979–1991) in the city centers of Phnom Penh and Battambang after the fall of the Khmer Rouge. It tells the stories of state workers who repaired and reformulated media infrastructure within a context of many obstacles, including extreme poverty and continued but often unpredictable interparty violence. This work was deeply political, as film and radio in this period remained central to the political workings of the state and its wars with opposition parties. The media workers I interviewed came from a particular position as state workers. The state also enacted censorship, and Eastern Bloc countries contributed heavy-handed media advising. Noting this context, I argue that media workers demonstrated agility and resourcefulness in rebuilding after disaster. This work was highly affective and the media workers were recovering from losses and acute trauma from the Khmer Rouge.

The third chapter describes the violence and cultural frictions that emerged in the process of (re)constructing the material, social, and ideological infrastructures of independent media in 1990s Cambodia. Under Cold War conditions, the United Nations backed a coalition that included the Khmer Rouge until the Paris Peace Accords in October 1991, when borders officially opened to Western trade. In early 1992, over 20,000 foreign personnel entered the country as part of the United Nations Transitional Authority Cambodia (UNTAC) to run an election in May 1993. The third chapter argues that UNTAC worked in the Cambodian media sector naive to the historical links between the media sector and political power, instigating further violence. As an ahistorical form of work, I contrast UNTAC's media work to infrastructural restitution and, in so doing, illuminate some of the dangers of ahistorical work and the conservative elements of infrastructural restitution. The first case in this chapter concerns the construction of the Radio UNTAC broadcasting station, the first nonstate media outlet since the Khmer Rouge came to power. The second half of the chapter describes broader media transitions ushered in by opening to Western trade and economic privatization during this period.

In part II, chapter 4 focuses on infrastructural restitution's quality as a subtle form of political action. It gives the case of Roung Kon, a group of architects who find, survey, and exhibit the models of heritage cinemas of Cambodia, built before the war. Building on the theory of ruins, rubble, and hauntings, I give a sense of the complex and contradictory affective resonance of cinemas as what I call *media ruins*. They represent modernist and technological space in decay, inhabited by memories and ghosts, including the ghostly presence of movies themselves. The infrastructural restitution of media ruins acts as a form of hidden script of political action in an increasingly authoritarian context in contemporary Phnom Penh. Roung Kon praises and mourns the flourishing arts and space of the cinema from the Sangkum Reastr Niyum partly to encourage public support for the arts and sustainable urban development. In so doing, they suggest an alternative to contemporary information control.

The fifth chapter illustrates how infrastructural restitution is way to access positive affect about the national past and, as such, works as a kind of healing action. The chapter describes a contemporary Cambodian youth group called Preah Sorya (Sun God) who research, recover, distribute, and screen Cambodian films from before the Khmer Rouge period. This chapter focuses on the relationship between trauma and media materiality. Preah Sorya does their work in order to commemorate those lost during the war period, heal what they call "problems of the way of the heart" (*banhhaa phlauv chett*), and dream toward new futures. They creatively incorporate *disintegration noise* (the decay of film or overlapping of multiple historical formats) into their live-action and online projects.

The sixth and final chapter returns to the transnational tensions of infrastructural restitution. It analyzes and contrasts the creation, use, and promotion of two internet-based platforms (Amazing Cambodia and the Bophana Center's App-learning on Khmer Rouge History) that store and disseminate historical photographs, music, film clips, and historical documents, using Facebook and a smartphone application. This chapter describes the rural-urban mobility of infrastructural restitution, as both projects involve substantial research and dissemination trips around the Cambodian provinces. It focuses on the interplay of influence between the US-based tech companies, foreign donor influence, and the grassroots, locally oriented practices of the creators of the tools.

I MEDIA HISTORY

1
INFRASTRUCTURAL HISTORY: AUDIOVISUAL MEDIA IN THE SANGKUM REASTR NIYUM (1955–1970)

In *Rose of Bokor* (1969), Norodom Sihanouk, then the head of state of Cambodia, plays a Japanese general who, in 1945, took over Bokor Mountain, a Cambodian hill town and French colonial base.[1] The film depicts Cambodian villagers declaring independence shortly after the new occupation. Meanwhile, Sihnaouk's character falls in love with Rose, a wealthy Cambodian female farmer, played by Sihanouk's wife Monique Izzi. Sihanouk's character tells Rose that he has "always loved the arts more than the martial arts." Through the film, Sihanouk establishes his version of Cambodia: an idyllic place with glorious natural beauty and the wonders of Angkor but one that always seems to be stuck in the middle of troublesome international politics. In the film, the Japanese ultimately lose the war, the French crush the Cambodian independence movement, and Sihanouk's character commits suicide. Lady Monique's character concludes the film by saying, "What a terrible thing, we are very unlucky."

One of the most puzzling things about *Rose of Bokor* is Sihanouk's very attention to it in late 1969. Why would Sihanouk, head of state of Cambodia, focus on making a historical entertainment film rather than addressing the mounting pressures coming from both inside and outside his country? To answer this question, in this chapter, I analyze the push and pull between Sihanouk's regime and foreign powers, particularly the United States, in building audiovisual media infrastructure in the early postcolonial period, and how these technologies circulated within Cambodia. I show audiovisual media have an air of magic and a power of persuasiveness that have given them a central place in the lives and politics of Cambodian political elites, urban and rural poor, and foreign powers for the last sixty-five years.

My first argument in this chapter is a straightforward historical type. I argue that the investment of the USIS in film and radio between 1955 and 1963 strengthened and supported national audiovisual media infrastructures that then were deployed to undergird Sihanouk's state power. The USIS stationed in Phnom Penh intervened in Cambodian domestic affairs by importing radio and film technologies, training Cambodian Ministry of Information staff about the use and repair of equipment, and disseminating film and radio widely through provinces. The newly independent state at first accepted these audiovisual infrastructures as a part of the goal to improve national media. Struggles for control over film and radio infrastructures, however, were a significant factor in the disintegration of US-Cambodian relations at the end of 1963 (finalized in 1965). In the 1964–1970 period, these media infrastructures became critical elements in Sihanouk's strategy for authoritarian rule. Sihanouk and his staff used audiovisual media equipment as tools for state power. They tightly controlled radio and film technologies as they became increasingly important parts of his ruling strategy.

My second argument concerns the value of an infrastructural methodology and analysis in a history of media, which opens up media history to insights around materiality, work, and relationality.[2] I analyze early postcolonial media with a particular eye to its history of devices: radios, transmitters, boats, screens, and reels, emphasizing that media can be understood not only through content but also through form.[3] Media experiences during this period were largely ephemeral and existed in radio waves and transient mobile cinemas. I perused the archival record for images and descriptions of devices in order to recreate these scenes and give a sense of the material conditions under which independent media creators could produce new content and the ways that average Cambodians gained access to media forms.

When I move to stories of reconstruction in later chapters, the media products that most of the actors restore come from the Sangkum Reastr Niyum period, now often described as a cultural "golden age." This chapter gives the critical context for this reconstruction and shows that the Sangkum Reastr Niyum period might not have been as rosy as later media producers often make it out to be, who see it through the lens of what happened later, that is, the devastation of the Khmer Rouge. This period, in reality, involved vast amounts of inequality, foreign interference, and authoritarian rule, making it an economic and geopolitical context not wholly dissimilar to Cambodia in 2017. The media creators who made innovative art during this period did so often by co-opting or working around these conditions in a way that foreshadows contemporary media production.

In this chapter, the components of media infrastructure that I pay closest attention to include media spaces (areas of collective listening and watching), the training of media specialists, the travel of technicians in and out of the country, and the movement of film and radio equipment from foreign countries to Cambodia (and sometimes back) and from urban to rural areas within Cambodia. I often describe in some detail the material qualities of media—specifics of technical models, where technologies came from, how they spread around Cambodia, and how they deteriorated in Cambodia—to show that transnational and national negotiations happened by and with material things as well as with ideas. I also consider the legal regulations for media; for example, in what moments and under what conditions were certain foreign influences or internal dissonance accepted or violently cut off?

Taking this approach, I focus less on the content of media and the give-and-takes between media content and social norms than some histories of media do. An infrastructural media history provides a new set of insights about the role of media in society and politics. It emphasizes the relationality, transferability, and movement of media and the tangible and finite materiality of that circulation. It allows us to see the conditions of possibility for independent artists to appropriate media tools (often first used by imperial or authoritarian powers) in new ways, or for people far from centers of power and wealth to appreciate the products of these tools. Infrastructure contributed to both the common experience and mechanisms for control of media during this period.

I proceed in three sections. In the first section, I move into my historical argumentation and I tell the radio and film history of Cambodia from the end of the colonial period through the breaking of economic ties with the United States in late 1963. In the second section, I describe radio infrastructure in 1963 to 1970, focusing on Sihanouk's attempt to control radio waves in remote and border regions of the country. In the third section, I focus on film infrastructure from 1963 to 1970, outlining the ways that national cinema, foreign technologies and techniques, and independent filmmakers intersected and how Sihanouk manipulated media through film censorship and investment in his own films. I conclude by revealing how the history of radio and film infrastructures in Cambodia opened up an environment for independent radio and film artists to create innovative content within the bounds of Sihanouk's regime.

My sources in this chapter primarily come from the National Museum of Cambodia and the National Archives of Cambodia. At the National Museum, I reviewed the papers of Ingrid Muan, an American art historian who made copies of many USIS documents from the US National Archive and tragically died in 2004. I reviewed all documentation in the National Archives of Cambodia about radio, television, and cinema, their collection of telecom documents (mostly government documents and speeches), and their entire collection of *Réalités cambodgiennes*, a popular French-language weekly magazine that ran during the Sangkum period and was overseen by the Sihanouk government (1965–1970).[4] I also reviewed scholarship from the late 1950s Cambodia, focusing particularly on descriptions of media use. I scanned historical documents with a particular eye to seeking out descriptions and images of media's form (rather than simply media content); for example, I looked for descriptions and mentions of electronic devices, towers, and cables and images of televisions, mobile cinemas, and radios. These often had to be dug for and were often found within news clippings, in memoirs, in advertisements, or in magazines. Sihanouk's films themselves (such as the *Rose of Bokor*) also played a role as primary sources.

The challenges of any historical research in Cambodia are nontrivial; the archival record in (nearly) all of the archives I visited was scattered. During the Khmer Rouge period, many documents were neglected or targeted for destruction. Sometimes I fell upon certain collections by chance or through an unusual search term; I also got tips for key collections from fellow researchers.[5] In addition to archival research, I also found some primary materials online. The Southeast Asia digital library, for example, has visual repositories including May Ebihara's images.

This is a story constructed largely by elite sources (foreign or domestic), but I argue that by taking the infrastructural approach described above, we can still understand the ways that media mattered to many people in Cambodia during this period. Cambodia has many archival gaps due to the destruction of sources by the Khmer Rouge regime and low literacy levels during the Cold War period, limiting the availability of documentation produced about, let alone by, many Cambodians. Undoubtedly, media had a limited audience due to cost prohibitions for products like radios and batteries; environmental destruction of equipment from heat, dust, and humidity; lack of electricity in large sections of the country; failures of language translation; and the urban location of cinemas.

These conditions also led to specific radio and film cultures in Cambodia. For example, because of Cambodia's economic limitations in the Cold War period, USIS made mobile cinemas; villagers developed the practice of sharing radios; loudspeakers in key public spaces enabled larger numbers to hear broadcasts in provincial towns; and other accommodations were made. As a consequence, audiovisual media reached many Cambodians in the countryside and in provincial towns. Outlining this history below helps us get a sense of how media technologies were received, thought about, and used in innovative ways. It also provides context for why media technologies were seen as so important to control of the country. This history also gives us a deeper sense of the constraints and possibilities for Cambodian radio artists and filmmakers during this period (particularly after 1963), and shows the conditions under which they were able to create new kinds of films and radio content, despite the censorship and control of the Sangkum Reastr Niyum period.

1 RADIO AND FILM INFRASTRUCTURES UNTIL 1963

Cambodia was part of French Indochina from 1863 to 1953, but audiovisual media were never a focus of French colonizers, who refused to develop its infrastructure until the World War II and early independence period.[6] The first large-scale radio project in Indochina began when the station "Radio Saigon" started in 1939 under the government direction of Governor-General Jean Decoux of Vichy France.[7] In that year, the French government also ruled that all French and Cambodians in Cambodia who wanted to own a radio needed to be registered. These licenses authorized people to have a radio for receiving communications but did not allow them to transmit correspondence.[8] Radio Saigon broadcast in French, which meant it had limited relevance to Khmer-speaking people in Cambodia. As in other sectors of the colonial government, the French invested much more in Vietnamese than Cambodian radio infrastructure, and Radio Saigon covered issues related to Vietnam but rarely Cambodia.[9]

In 1941, during World War II and early Cambodian struggles for independence, King Sisowath Monivong died. Though Monivong's son Sisowath

Monireth was the heir to the throne, the French authorities chose his grandson, nineteen-year-old Norodom Sihanouk, to be Cambodia's next king because they expected him to be easier to work with.[10] Sihanouk was reportedly timid in the role at first. However, during the Japanese occupation in March 1945, Sihanouk briefly declared Cambodia independent, changing its name from Cambodge to Kampuchea and invalidating Franco-Cambodian agreements (as dramatized by the villagers rising up in *Rose of Bokor*). But in October 1945, after the end of the war, Sihanouk reopened negotiations with the French and signed an (albeit weaker) modus vivendi to reestablish Cambodia as part of French Indochina in early 1946.

This moment marked the beginning of the phaseout of French rule and the growth of the political role of radio in Indochina. The French realized the potential threat of independent radio voices and developed a newfound interest and investment in radio infrastructure.[11] This French investment, however, came too late and was ultimately unsuccessful. In 1949, Sihanouk negotiated an agreement in which Cambodia received some autonomy for military and foreign affairs, which he called gaining "fifty percent" of Cambodia's independence. On April 13, 1950, Radio Saigon became Radio France-Asia in Saigon, and other radio broadcasting in Indochina became the responsibility of the individual states. In July 1950, the Cambodian government took over "Radio Cambodge" in Phnom Penh from French management.[12]

Along with taking over control of the national radio, Sihanouk also started a national film unit. Sihanouk was reportedly interested in film from an early age when his parents taught him to "love romanticism," including the cinema.[13] He created the Office of Film within the Ministry of Information in 1951, during the transition from French to Cambodian rule. At this time there was already a vibrant cinema culture in urban Cambodia, with thirteen cinemas total in the country (ten of them in Phnom Penh, three in provincial capitals). French investors financed eight of the theaters and Chinese investors financed the other five. Each had a 35-mm projector and ten also had a 16-mm projector imported from outside the country.[14] Their total seating capacity was over 5,000, with a combined annual audience estimated at 1.5 million; the cinemas were reported "to be packed at almost every showing."[15] Programs usually consisted of a feature, a newsreel, and a documentary or a cartoon.[16] Films were all imported, however, and were in foreign languages. Some had Cambodian subtitles, but since many Cambodians did not know how to read, these were of limited utility. Some films had live translators at the front of the theater as was common in the Southeast Asian region at this time.[17] Despite some difficulties understanding the language of the films, they exposed Cambodians to scenes of life in places like India, the United States, and Hong Kong.

Sihanouk recognized the power of media and used it not just for entertainment but also for persuasion. His negotiation for independence in 1953 was dramatic, and it was the first time that he demonstrated his abilities to use and manipulate media for his own ends. In January 1953, Sihanouk dissolved the National Assembly and declared martial law. He traveled to France, saying he needed to visit the country "for his health." He wrote to the French president, Vincent Auriol, and argued that though *he* was loyal to France, he could not guarantee the loyalty of his citizens. Auriol told him to go home. On the way back to Cambodia, Sihanouk traveled to and gave radio interviews to stations in Canada, the United States, and Japan, criticizing the French. He continued to travel internationally and criticize the French for the rest of the year on international television, radio, and newspapers until they essentially gave into his demands, granting Cambodia full independence in November 1953.[18]

During the period of Sihanouk's fight for and then establishment of independence, the United States became actively involved in Indochinese politics and audiovisual media infrastructure. The USIS began in the early days of the Cold War to work around the world to fight the threat of advancing communism with a cultural war promoting American media and creating local anticommunist propaganda.[19] Southeast Asia, especially Vietnam, was an early site for this new war and the USIS started to invest heavily in film, radio, and other propaganda materials to "win hearts and minds" for the anticommunist cause.[20]

Beginning in 1950, the USIS took advantage of the limited media infrastructures and the weak involvement of the French in Indochina as an opportunity for the United States to build its influence and fight the communist influence there through audiovisual channels.[21] Between 1950 and 1955, the USIS started supplying goods to the Cambodian government, including mobile units for disseminating films, a printing plant, a photographic laboratory, a radio transmitter and receivers, tape recorders, and public address systems.[22] Like the French during the colonial era, the Americans at first focused their work in Vietnam, making Saigon their base. As of November 15, 1950, Radio Cambodge started to rebroadcast the USIS-sponsored Voice of America (VOA) programs in English, French, and Vietnamese—but not Khmer.[23] A 1950 USIS telegram from Saigon to Washington explained that Cambodians were upset at how the VOA sometimes confused the differences between Cambodia and Vietnam.[24]

The political landscape changed in early 1955, when Sihanouk abdicated the throne and reentered the political realm as a private citizen to run for president as the representative of his party "Sangkum Reastr Niyum" or "community of the common people." His father Norodom Suramit took over the role of king. Sihanouk rigged the September 1955 general election by, among other forms of intimidation, shutting down opposition newspapers. Unsurprisingly, given the questionable legitimacy of the election, Sihanouk's party officially won in 1955.[25] Between 1955 and 1970, Sihanouk and his party ruled in an authoritarian fashion. His party was based on conservative social values, was pro-nationalist and pro-monarchy, and integrated Theravada Buddhist teachings. Sihanouk's one-party rule was tightly controlled and dissidents were not tolerated. He encouraged the wealthy to give money to poor people as a way to gain Buddhist merit. Society acted in practice like cronyism; Sihanouk set up state enterprises and then allowed politically aligned elite to manage it, often for their own personal gain. Foreign advisors courted the government with expensive gifts. All civil servants were required to demonstrate their "loyalty" to Sihanouk through rituals of membership to the royal party; these included supervised participation in a few weeks of manual labor each year, and appearances in parades for Sihanouk. There existed an amazing discrepancy of wealth.[26]

In early 1955, just after the establishment of the Sangkum Reastr Niyum government, American radio and film technicians started working out of a Phnom Penh office. In October 1954, the USIS completed an evaluation of Cambodian radio infrastructure, reporting that it was weak and calling for more American intervention. The country had 4,800 receivers for a country of five million (approximately one radio receiver for every 1,000 people). The two transmitters in Cambodia were also in poor condition, the primary one in Phnom Penh needing repairs and the second in Battambang nearly inoperable. The number of goods the United States supplied to the Cambodian Ministry of Information increased dramatically. In 1955–1956 the USIS donated a transmitter, over 1,200 receivers, and "a quantity" of public address systems (speakers for playing radio in market squares and other places).[27] The American mission tried to distribute radios to public places such as meeting halls, markets, and pagodas.[28] The radios needed electricity, so the United States also supplied 150 generators where no electricity was available.

The USIS also trained Ministry of Information staff starting in 1955–1956. They taught them how to use new radio receivers and transmitters and repair those distributed earlier in the decade.[29] They also sought to help the Cambodian government expand and improve radio programming, for example, by aiding the Education Division of the Ministry of Information to script and produce educational programs to supplement the news and music that they already played.[30]

On August 15, 1955, VOA also started broadcasting in the Khmer language from Phnom Penh. The broadcast, recorded in Washington, DC, consisted of fifteen minutes of Cambodian broadcast news and fifteen minutes of commentary and features in the morning on national radio frequencies.[31] President Dwight D. Eisenhower and King Norodom Suramit wrote telegrams to each other to celebrate the beginning of VOA in Khmer and what it meant about the relationship between the two countries.[32] Beneath the surface, however, the Cambodian government worried about the VOA as a mechanism of propaganda. They asked for Cambodian embassy oversight of the VOA programming, and the United States offered to coordinate with the Cambodian Ministry of Information about broadcasting to ensure that the broadcast was a "meeting of minds" and a sign of friendship.[33]

The VOA quickly developed a large listening audience, and the USIS designed programs to change the radio status quo, within the bounds of Sihnaouk's neutral state. USIS staff wrote in a memo that Radio Cambodge didn't "stimulate" listeners and that the radio simply "told" them, without controversial or provocative information. The VOA wanted to change this standard and add more "stimulating" (i.e., political and interactive) content.[34] The USIS based in Phnom Penh promised a coronation photo of the Cambodian royal family to anyone who wrote a letter to the station as a first step at making the station more interactive. They noticed with confusion that letters were not stamped and sent by postal service but instead were personally carried to the station, likely because there was not an actively used postal service in the country at that time.[35] This is one example of how the USIS, including the VOA, tried to construct new media environments without fully understanding how society worked in Cambodia.

By 1956, the USIS program also started helping to develop Cambodian film production and dissemination programs. In April 1956, the USIS made the first local film in Cambodia, in Khmer.[36] By November of that year, the USIS film section completed seventeen newsreels and seven documentaries with Khmer and English soundtracks. They also gave thirty-five 16-mm projectors to the Cambodian government for their own information purposes.[37]

The USIS built an infrastructure of film mobility and expanded the dissemination of cinema to remote locations. They planned elaborate public showings and moved films, projectors, and screens around by what they called "cinecars" actively in 1956 and 1957.[38] Evaluations reported that 150,000 people every month watched USIS films via mobile units by the beginning of 1957. The USIS was particularly proud of the diverse rural audiences they were able to attract.[39] In fact, one USIS officer claimed that interest increased proportionally with the lack of social and economic development in an area.[40] The USIS trained Cambodian staff to run the cinecar, repair equipment, and act as cameramen, editors, and sound technicians.[41]

By late 1956, the USIS reached more rural populations by distributing films by boat, reaching populations that were inaccessible by road, particularly during the rainy season, and that lived on the Sangker River (which connects the Cardamom Mountains to Tonle Sap lake). During the rainy season of 1956, they offered seventeen film showings to villages along the banks of the river, reaching a total of 8,500 people. In fifteen of the seventeen trips, it was the first time that the townspeople had ever seen movies. The boat traveled every week from Monday to Saturday and was (according to American reports) enormously successful, improving the population's goodwill toward the USIS. The report states, "The USIS boat received an excellent reception at every village. At most villages, the local chief would be waiting the arrival of the boat with many eager hands willing to assist in unloading equipment."[42] The cinecars and boats expanded the media geography extensively. Building a new mobile film infrastructure this period allowed the USIS and the Cambodian government to control messages sent to rural Cambodians who were largely cut off from other global influences. These films became an important mechanism for spreading the anticommunist message.

In 1956, the USIS also built cultural centers in Battambang, Siem Reap, and five other provincial capitals where radio could be broadcast and films could be screened. These centers were "constructed by joint effort" between the Cambodian government and the USIS.[43] The centers played the VOA and other radio programs every day and arranged for public film screenings regularly. The cultural centers provided workspaces for audiovisual creation and repair to radios and other audiovisual equipment that had been previously distributed. The centers also arranged showings of USIS films in "outlying localities."[44] In November and December 1956, there were hundreds to thousands of attendees per show.[45] In a similar model, the USIS aided the Cambodian government with their own "information halls" for distributing the news around the country. They supplied them all with radios and film projectors in order to improve communication channels between the Cambodian government and people in the provinces.[46]

The most popular USIS films during this time were *Boy Scout Jamboree*, a color film, and *Defense against Invasion*, an animated public health film.[47] Other films focused on American lifestyles and included such titles as *Life in America* and *Buddhist Art in America*.[48] The USIS reported that films depicting athletic events, youth activities, and animation were most popular, especially those in color. Surprisingly, the USIS found that the film *Cambodian Coronation* was unpopular, which they attributed to peasants remote from the city being largely uninterested in Phnom Penh politics.[49]

Though these films promoted American values and lifestyles, they were rarely explicitly political. USIS films needed to remain politically subtle for the USIS to stay in Sihanouk's favor. Even in this time of general alliance between the United States and Cambodia, Sihanouk was still skeptical about US intentions. On September 22, 1957, Sihanouk reiterated to the USIS that "propaganda was not allowed" and reminded the unit that their material could not be overtly political. The USIS filmmakers agreed that they would focus not on politics but instead on a wide range of films about public health, American culture, and Cambodian culture. For the time being, this justification was accepted.[50] The compromises that the American team made for Sihanouk's demands served to make the primary outcome of these projects the support and strengthening of Cambodian national audiovisual infrastructure, rather than promoting anticommunist propaganda.

At this time, USIS activities were interpreted in a number of ways by Cambodian audiences. According to Steinberg, a social scientist studying

Cambodia in 1957, many rural Cambodians appreciated the majority of USIS media. USIS self-evaluations reported that remote villages found the USIS shows exciting because these were often the first films villagers had seen. The USIS was also the first organization to create original Khmer-language film and broadened the scope of Khmer-language radio broadcasting. They therefore reached a far larger audience directly with media, without the filter of a foreign language, subtitles, or the occasional live translator. Steinberg also noted some criticisms of the USIS activities. He said that when military messages were shown along cultural products side they could be seen as "warmongering."[51] Some Cambodians also worried that the United States was "simply trying to take the place recently occupied by the French."[52]

By 1958, the national radio station—Radiodiffusion National Khmère, or Radio Cambodge—had matured under the guidance of USIS trainings and donations. It broadcasted programs in Khmer, Chinese, Thai, Vietnamese, English, and French, including sports, the national lottery, stories of agriculture, weather, dance, music, and "muscular awakening." In 1957, 7,000 receivers were registered.[53]

Despite the strong investment and interest in the radio from both domestic and international powers, a question remains about how many people were accessing and listening to radio, particularly in rural Cambodia in the late 1950s. May Ebihara spent a year from 1959 to 1960 in Svay, a small rural town in Kandal Province. Based on her ethnographic experience there, she wrote a remarkably detailed picture of village life in Cambodia.[54] She brought a radio with her and wrote, "My house became a place to visit for company or curiosity, to listen to the radio, and to receive simple medications."[55] She gave an overview of how news was generally received in her village:

> News of national and international politics does reach the village in several ways, although Svay residents have relatively little access to most media and are often uninterested in political affairs. There are only two radios in the entire village (both located in the other hamlets), although I brought one into West Svay and there are others at the local temples and in Kompong Kantuot. The Khmer Broadcasting System (Radio Diffusion Khmere), operated by the government offers music, drama, news, and speeches on its one station. As a source of news the radio is of limited utility to villagers because the news broadcasts are either in French or in the formal Khmer speech used by educated people that is barely intelligible to most peasants. Whenever Sihanouk is on the radio, he arouses great interest and attention (and it is a credit to

Figure 1.1
A crowd listening to May Ebihara's radio in Svay, 1959. Source: Southeast Asian Digital Collection, Northern Illinois University

> Sihanouk's cleverness that, in such speeches to the populace, he always uses "colloquial" language in at least part of his talk so that ordinary people can understand him). Otherwise, however, West Svay villagers are much more interested in listening to music and especially dramas (which always drew great crowds around the radio).

As Ebihara pointed out, even with Khmer-language media materials, formal language could exclude many peasants from understanding and engaging with radio and film. Language issues were not the only barrier to receiving radio news. Steinberg explained that "inadequacy of electric power seriously restricts the use of receiving sets away from the large towns."[56] However, Steinberg continued, "Given the still small number of radio sets actually available . . . every owner of a radio set keeps his whole neighborhood informed of the latest news. . . . In certain small towns where the mayor owns a set, a public address system may be connected with it, so that people by collecting in the market square may listen to news and entertainment." He explained that public address systems were sometimes also set up in pagodas.[57] Some villagers could not get away from the public address of radio even if they wanted to. Steinberg wrote, "Loud speakers are considered the most effective means of persuading groups of people in villages or cities. There appear to be no taboos affecting the location of mobile units in Cambodian villages, and operators are understood to get the full cooperation of local officials."[58]

The growth and circulation of radio and film infrastructure—developed in part by the United States—continued as the country became a more mature independent nation in the late 1950s. Media relations began to change, however, as Cambodian-US diplomacy deteriorated after the so-called Dap Chhuon affair, which, I argue, was centrally about radio. In February 1959, Dap Chhuon, a right-wing warlord who exercised a large military influence over Siem Reap province, executed a plot to overthrow Sihanouk and install a Western-friendly government. He had formerly been an anti-French Issarak rebel and had amassed weapons and an army of about 3,000 men. He also had the backing of 1,200 Khmer Serei, a South Vietnamese-based anticommunist rebel group with fighters in Thailand. The government learned of the plot and took over Chhuon's villa, where they found gold, radio equipment, and two Southern Vietnamese technicians. Chhuon was executed a few days after he was found.[59] The Khmer Serei continued to operate from South Vietnam and the Thai border areas. Sihanouk was rightfully suspicious that the Americans were in collaboration with the Southern Vietnamese supporting the Khmer Serei. A US State Department official and former soldier named Victor Matsui provided the radio equipment found in Dap Chhuon's house in 1959.[60] Though the United States has never officially admitted to assisting the coup and argues that the radio was provided "only to keep tabs on Chhuon's scheming," subsequent evidence strongly indicates that the United States worked with the Khmer Serei to overthrow Sihanouk in this plot and that Sihanouk's anger was therefore justified. In 1967, Sihanouk made a film called *Shadow over Angkor* about the Dap Chhuon affair in which he plays the commander who captures Dap Chhuon.

As Cambodian-US relations were deteriorating, Sihanouk turned to China for more support. The United States collated a "Chinese affairs summary" on a monthly basis that concluded that communists had "radio, motion pictures, publications [in Cambodia] . . . but less than the US."[61] Chinese media were particularly geared toward people of Chinese and Vietnamese ethnicity who lived in Cambodia.[62] Smaller communist states also were involved in Cambodian media; for example, Hungary started a radio broadcast in Cambodia in November 1957.[63]

The most significant investment in Cambodian media infrastructure from a communist state was the Chinese donation of a major radio transmitter for Radio Cambodge built in Stung Meanchey, in the outskirts of Phnom Penh. Relations between Cambodia and China warmed in the mid-1950s, when Sihanouk first met Chinese Premiere Zhou Enlai at the Bandung Conference in April 1955. There leaders of newly independent former colonies in Asia and Africa gathered to discuss how to support decolonization. Between 1957 and 1962, the Chinese financed and technically supervised the building of the state of Cambodia's new radio transmitter and an accompanying studio facility. In 1957, two Chinese engineers came to Cambodia to assess the project. In July 1958, Beijing and Phnom Penh established official diplomatic relations, and by November 1958, Sihanouk laid the first foundation stone in the new transmitter building.[64] The first test of the transmitter occurred on April 14, 1959. On January 12, 1960, the Chinese hosted a ceremony to hand over the first installment of work.[65] In this ceremony, the Chinese ambassador explained that the "The Royal Khmer National Radio Broadcasting Station has begun transmitting its waves around the world, a station symbolizing friendship and collaboration between China and Cambodia."[66] The transmitter provided 20 kW on medium wave and another 15 kW on shortwave, which represented a major improvement from the weaker, earlier transmitter donated by the US (10 kW medium wave).[67] This greater strength of transmission led to greater coverage of the radio waves across the country. Between 1960 and 1962, the Chinese engineers upgraded the emitter to a 50-kW signal and built an auditorium. Chen Shu Liang, the Chinese ambassador to Cambodia, congratulated Cambodia on its recent modernization and development: "In just a few years, modern factories, roads and institutions of popular charity are emerging in every corner of the Kingdom. The Emission Station is like a flower in a blooming garden, a flower offered by the Prime Minister Chou En Lai to Prince Norodom Sihanouk, in mark of the Chinese people's friendship."[68]

Sihanouk was at this time maintaining Cambodia's neutrality, but both the Chinese and the Americans were beginning to push him harder to take a position on Cold War politics. His earlier warmth to American USIS initiatives began to cool. In 1961, US Embassy staff reported that Sihanouk was increasingly hesitant about American cultural centers. He remarked that if there is an American-Cambodian cultural center, "there might also be a Soviet/Chinese cultural center."[69] But between 1961 and 1963, US aid still accounted for over 14 percent of Cambodian governmental revenues, including 30 percent of the military budget.[70] Relations finally took a nosedive in August 1963, when a new Khmer Serei radio program again enraged Sihanouk.[71] This radio station broadcast from Southern Vietnam and could be heard in southern regions of Cambodia. An US news article from 1964 explains,

> The prince made it plain at that time that he regarded the broadcasts as an indirect American attack upon him. As the broadcasts continued, despite his protests, the Prince's public displeasure with the US grew. Finally, on November 5, 1963, Prince Sihanouk issued an ultimatum: all US aid activities in Cambodia were to be stopped unless the Khmer Serei radio was silenced. Though December 31 was the deadline he set, the Prince's patience in the face of the clandestine radio attacks gave out in December and he ordered the American military and economic aid missions to pack up and leave.[72]

Relations continued to deteriorate until April 1965 when Sihnaouk severed diplomatic relations. The embassy was closed on May 3, 1965.[73]

There was some concern among the USIS staff even during its peak activity that all the equipment that the United States was bringing into the country could be used for ends that were not intended. One report noted that "it has been informally argued that mobile units may fall into hands of communists, or otherwise used counter to the US objective of assisting Cambodia to build a . . . US assisted audiovisual program in Cambodia—itself a project to help the government strengthen (and in some cases, create) vital links of communication between itself and its people."[74] The program continued despite these concerns. Media infrastructures did in fact enable Sihanouk's authoritarian rule, as the United States feared, and as explicated in the next section. However, independent Cambodian artists and technologists were also able to appropriate these audiovisual infrastructures creatively to make innovative forms of content within the constraints of the Sangkum Reastr Niyum state.

2 SANGKUM REASTR NIYUM RADIO INFRASTRUCTURE

Sihanouk believed strongly that foreign radio waves must be defended against and the strength of the Phnom Penh signal protected. Cambodia, a small country with more powerful neighbors, was vulnerable to stronger radio transmissions. Sihanouk became obsessed with the strength of the national radio station's transmission, radio waves coming from across the Southern Vietnamese and Thai borders, and ways to strengthen and promote the Cambodian radio in rural and border regions. Alarmed authors wrote articles about radio infrastructure that regularly appeared in the Sihanouk-sponsored publication *Réalités*, suggesting that the government cared deeply about both developing national radio infrastructure and controlling foreign broadcasts in Cambodia.

In early January 1967, *Réalités* reported with concern that the "Voice of Free Asia" was broadcasting in rural parts of Cambodia. The article explained the United States built a 1,000-kW transmitter (compared to the 20-kW transmitter that the Chinese built at Stung Mencheay) in Thailand to broadcast the Voice of Free Asia in Chinese, Malay, Indonesian, Khmer, and Lao for eight hours a day.[75] The article continued,

> Let's face it. It is certain, given the exceptional power of this transmitter, that it will be heard perfectly in Cambodia. The Khmer people are curious by nature and, while they are suspicious of what comes from Thailand and the USA, they will not fail to listen to a station so easily audible that speaks to them in Khmer. If the Americans do not make the irreparable mistake of entrusting the Cambodian emission to the Khmers Serei for their propaganda of hatred, if they give this emission a relatively "objective" look, it will be heard on the border and at the seashore. [In these places] Radio Phnom Penh is difficult (and sometimes impossible) to hear if only a modest medium-wave transistor is available.[76]

To address these concerns, a royal engineer visited all the regions to measure the power of the Radio Cambodge broadcast.

Later in January 1967, *Réalités* reported that the Ministry of Information was able to start using a new transmitter, which ran on a new frequency in addition to the current frequency.[77] The Ministry would use it to play Radio Cambodge over the same frequency as the international broadcasts, hoping that the quality of sound for the national radio would be better than the international radio. A few weeks later, a *Réalités* author wrote, "A solution is necessary for the urgent problem of the international radio transmissions." He suggested that the government needed to build new relay transmitters either alone or in cooperation with a "friendly country" to distribute the radio more broadly. A relay tower would receive a signal from the Phnom

Penh transmitter, amplify it, and then retransmit it to the nearby region. The article's author suggests that Cambodia should build relays in Bokor, Battambang, Svay Rieng, and Kratie to cover the whole of the country. If the Phnom Penh radio station were stronger and more audible, then the border populations would be less likely to listen to foreign broadcasts and would listen to ones coming from the national radio.[78] Later in February 1967 *Réalités* reported that because of a new relay tower in Bokor, Radio Cambodge was heard better in Bokor, Kep, and Sihanoukville.[79]

In November 1967, *Réalités* reported that Cambodia should also make an emergency radio in Kirirom for military action. The author reported that "the concentration of all of the radio waves poses a serious problem: an incident or a sabotage would not allow the country to listen to the instructions of the government. It will be therefore useful to create a station from one to five kW shortwave in a place easy to protect [Kirirom]." The author proposed to use the "numerous talents" of the army to run the radio technically and also perform live music. The author explained that in "normal time," the station could be used for fun and pleasure and would be complementary to Radio Cambodge. It could also connect urban dwellers who went to Bokor for weekends to Phnom Penh news.[80]

Though Sihanouk cared deeply about the political role of radio, radio was primarily, as hinted at in this article, listened to for entertainment and pleasure. During this period, radio theater also became an important art form and many Cambodians fondly listened to the radio collectively, particularly to *lkhoan niyeay* (spoken theater) or *lkhoan ayai* (comedy theater), played on the national radio frequencies. Traditional musicians and voice actors together would sit in the Phnom Penh radio station studio to create and record new music and dramas, which mixed story lines with musical interludes.[81] Some of the dramas are still popular today. Though the radio was used for news and propaganda, the same machine, the same frequency, and the same infrastructure offered a variety of content, including the music and stories popular in communities of rural and urban Cambodia, from children to the elderly.

3 SANGKUM REASTR NIYUM FILM INFRASTRUCTURES

By the 1960s, the National Office of Film expanded to include a new group to produce newsreels and educational films. The group incentivized local film production by giving half of the 40 percent tax on tickets back to Cambodian film producers.[82] Sihanouk deeply cared about film, perhaps a trait he learned from his mother, Her Majesty Queen Kossamak, who, throughout Sihanouk's rule, hosted film nights at her residence with Cambodian and foreign films.[83]

Som Sam Al's biography illustrates the ways that independent Cambodian cinema, national cinema, former USIS influence, and other foreign training were closely tied to each other during this period. Som Sam Al was a key figure in building a national audiovisual media program throughout the Sihanouk era. He traveled to France from 1949 to 1955 where he studied in a technical school for industrial design and mechanics as well as a school of photography and cinematography and interned for a French television station and the cinematographic service of the French army.[84] He then returned to work for the Ministry of Information and FARK (the Cambodian royal army), where he was trained by USIS technicians in 1955–1956. In the late 1950s he made some short movies about the indigenous hill people's "Khmerization" for FARK, full of claims about the benefits of integration into mainstream Cambodian society. He participated in a USIS-sponsored study trip to Japan, the Philippines, and Hong Kong. In 1957, he left the Ministry of Information to make his own films with the help of a team compiled of Cambodians who had also been trained by FARK and the USIS.

In 1960, Som Sam Al made the first full-length Cambodian color feature, titled *Phka Rik Phka Ruy* (Blooming Flower, Withering Flower), adapted from the novel by Ieng Say.[85] He made other commercial films (*Sobennavong* and *The Road of Happiness*) and collaborated with the French director Marcel Camus in the movie *Bird of Paradise* (1962). Som Sam Al worked in the national film studio, stocked with international film equipment. They had, for example, both a 35-mm Arriflex camera and a 16-mm Paillard, both imported from France. They also had a projection room, projectors, a studio for filming, and many props needed for making "Khmer-style" films, such as ancient costumes and materials for making Angkor Wat scenes.[86]

Som Sam Al also directly supported Sihanouk to make his own films. Between 1966 until his overthrow in 1970, when civil unrest was at a peak, Sihanouk became passionate about making films. During these four years, Sihanouk devoted much of his energy to the production of films, which he scripted, directed, wrote the musical scores for, and starred in. Som Sam

Al managed all of Sihanouk's films and claimed that Sihanouk was a real "cinéaste" even if he was an amateur.[87] With Som Sam Al's help, Sihanouk made his first feature-length film in 1966 (*Apsara*); then he made (produced, directed, wrote, and often starred in) six other films before being deposed in 1970. These include *The Enchanted Forest* (1966–1967), *Shadow over Angkor* (1967), *The Little Prince* (1967), *The Joy of Life* (1968), *Twilight* (1969), and the *Rose of Bokor* (1969/1979). Sihanouk also hosted film festivals during this period. The first was held in 1968, when Sihanouk's *The Little Prince* won the top prize (the Golden Apsara award). The second was in 1969, when Sihanouk's *Twilight* won the top prize. Sihanouk brought to the cinema screen stories of contented Cambodian peasants and a prosperous elite. Sihanouk's films often oscillated between Angkorian cultural references and the romanticization of Phnom Penh modernism by highlighting buildings and monuments created by architect Van Molyvann, music written and performed by Sinn Sissamouth, and modern Khmer fashion.[88]

The Sangkum Reastr Niyum and Lon Nol periods are now often referred to as Cambodia's "golden age of cinema," referring to the nearly 400 movies that were produced in Cambodia in Khmer by Cambodian directors and starring Cambodian actors between 1960 and 1975. These films were shown in movie theaters constructed in Phnom Penh and provincial capitals (Kampot, Battambang, Kratie, and Kampong Cham).[89] Mobile cinema also continued, both sponsored by the government and through private companies.[90] Movie-going was a highly popular activity and "during the festival days or on public holidays, it was almost impossible to get a ticket for a film screening, if you did not buy one in advance."[91] People in the theaters were reportedly deeply involved in the movies they watched. "People shouted and laughed when they watched scary or funny movies. . . . At other times they would talk back to the people on the screen. If an actor and actress played a mean or evil character, they often would get upset with them."[92] Candy, popcorn, dry lotus, watermelon, pumpkin seed, sugar cane, ice cream, and bread were favorite snacks of the cinema audience. In order to attract the attention of the audience, cinema owners put up huge, hand-painted posters in front of the cinema.[93] The soundtracks written for the films were another draw to the cinema.[94]

Though this was a time of exciting creative output, independent filmmakers needed to tap into the national and foreign audiovisual media infrastructures that were crafted under Sihanouk's government. Independent filmmakers often had connections to the government, foreign filmmakers, or the USIS, who would provide their equipment or technical training. Sun Bun Ly, for example, who formed the first independent commercial movie production company, Neak Poan Productions, in the 1950s, was first trained under the USIS.[95] The most popular and critically acclaimed Cambodian feature films of the 1960s era came out of Ly Bun Yim's and Yvon Hem's private studios. Both of these filmmakers had transnational influences. Ly Bun Yim attributed the beginning of his career to Americans who handed out free cameras associated with the "Life of America" photo competition.[96] Yvon Hem attributes his start in filmmaking to the 1962 filming of Marcel Camus's *Bird of Paradise*.[97]

The independent Cambodian films made during this period were generally socially conservative and in line with the values of the Sangkum Reastr Niyum regime. They often told stories based on traditional folktales. They regularly portrayed conservative gender roles with a certain set of subservient tropes reserved for female characters.[98] Often films were violent; those with the best special effects were often most popular. These could include scenes of wonders and miracles: flying horses, man-made earthquakes, and thunder, giants, and witches.[99] These were generally not outwardly political films of resistance as was the case in other parts of the Third World during the Cold War (and which are now called "Third Cinema").[100] Neither, however, did they offer overtly political propaganda for the cause of the national government.[101] Muan and Daravuth claim that the Cambodian films tended to attract less educated laborers who were drawn into supernatural stories and folktales, whereas students and white-collar workers tended to prefer international films.[102]

After the 1963 economic break with the United States, Phnom Penh theaters still played foreign films, but these films were limited to the Communist Bloc. In April 1966, Cambodia hosted a Soviet film festival;[103] in October 1966, a group of Chinese filmmakers came to Phnom Penh to make sports documentaries;[104] in November 1966, a team of Czechoslovak cineastes came to Cambodia;[105] and that same month the government hosted both a Romanian film gala and a Russian film festival.[106]

Film equipment and processing continued to move back and forth between Cambodia and other countries throughout this period. Producers began developing color film in Phnom Penh only in the mid-1960s despite poor quality due to a lack of temperature control. Yim, however, still had his films produced in France in the 1960s and in Hong Kong in the 1970s.[107] In the late 1960s, Roeum Sophon (another Ministry of Information and USIS-trained filmmaker who had been working at the Ministry of Information with Sihanouk since the early 1950s) brought back a machine from France that allowed soundtracks to be recorded separately then attached to a film. Yim described this as a key moment in the Cambodian film industry, giving film a consistent soundtrack, rather than fickle live narrations.[108]

The independent filmmakers were working within a context of a highly controlling government where foreign media influences were increasingly suspicious. By the mid- to late 1960s, censorship law controlled the projection of foreign media. For instance, in May 1966, a censorship law prohibited the playing of foreign radio in public places including "restaurants, shops, hotel halls, etc." If anybody chose to listen or watch these at home they had to "ensure that they could not be heard from outside." The article continues, "As can be seen, individual liberty is respected but foreign propaganda (which manifests itself everywhere) is not entitled to be here."[109] In October 1967, a law decreed that foreign films could be projected only with authorization of the Minister of Education, the Minister of Information, and the Undersecretary of State to the Presidency of the Council of Ministers. To justify this new law, a *Réalités* article complained that "the ideological invasion, as we see, is becoming more and more difficult."[110]

In this chapter I have shown how the USIS's investment in film and radio between 1955 and 1963 in Cambodia strengthened and helped develop national audiovisual media infrastructures that then supported Sihanouk's authoritarian regime. Struggles for control over these infrastructures, however, were a major reason for the disintegration of US-Cambodian relations at the end of 1963. The American support of Khmer Serei radio infrastructure beginning from the Dap Chhuon affair triggered Sihanouk to break ties with the United States. After this souring of relations, Sihanouk invested heavily in developing and tightly controlling national media infrastructures, which contributed to his authoritarian rule.

I argue that the government and foreign powers tried to control audiovisual media largely through infrastructure, rather than exclusively through content. Media histories sometimes neglect to address the formative influence and crucial shaping power of infrastructure. Many factors disconnected the media messages, regardless of who controlled them, and the Cambodian audience: language differences, lack of electricity, geographical separation, poor transportation networks to cinemas, and the difficulty broadcasting radio waves to remote parts of the country. The USIS and the national government attempted to build infrastructure to bridge these divides. Infrastructural control over media was both material and ephemeral and included the importation of new techniques, tools, and towers; the facilitation of travel of people and goods; and the attempt to strengthen and interrupt radio waves across space. The USIS bridged gaps by bringing movies to remote villages by boat and car, building cultural centers in provincial capitals, and distributing radios and projectors widely. In the field of radio, Sihanouk sought to limit the waves coming across the border and strengthen national radio transmitters to make the voice of Radio Cambodge stronger than other voices. This control also took legal form; from 1964 to 1970, Sihanouk sought to limit foreign influence in national media production particularly from the Americans through new censorship laws that banned unlicensed projection of foreign films in public spaces.

Media infrastructure also led to a certain kind of media experience common to most Cambodians. This media experience was shared in public places, through the development of a public cultural center system (run by both Americans and later the national government), mobile screenings, and loudspeaker systems for radio. These infrastructures were imposing and hard to escape. Yet this public orientation toward media became a way for Cambodians to gain back control over media experiences; they often shared their personally owned technologies with their communities and families as a grassroots development of infrastructure.

This media history also gives us a deeper sense of the conditions under which Cambodian artists were able to create new kinds of films and radio programs during the Sangkum Reastr Niyum period. Radio artists developed a culture of *lkhoan niyeay*, the recording of stories and music in the radio studio, which villagers would listen to collectively in public places. Independent filmmakers created the "golden age of cinema" after establishing connections to national media infrastructures and foreign filmmakers and technologies. As we will see in later chapters, these conditions foreshadow the ways that contemporary media creators, too, need to appropriate technologies and skills from foreign and authoritarian modes in order to create new content.

INTERLUDE: WARTIME MEDIA

Sihanouk worked hard to control the audiovisual media infrastructures that he knew were vital to his authoritarian regime. Yet by the late 1960s, foreign and internal pressures became too much for Sihanouk to successfully control the country, let alone the media. In July 1969, Sihanouk reestablished ties with the United States, but this strategic move came too late to prevent the United States from backing Sihanouk's right-wing military general Lon Nol, who deposed Sihanouk in March 1970 while he was away in Paris for medical treatment. The United States went on to heavily support Lon Nol's Khmer Republic in its civil war against the Khmer Rouge from 1970 to 1975.

The Lon Nol regime used the Phnom Penh radio station and the Stung Menchey transmitter first built during the Sihanouk regime. Many of my participants tell me that filmgoing continued to be popular in Phnom Penh through the civil war, even when bombs started going off in cinemas. In January 1975, a clandestine radio broadcast from the Khmer Rouge started operating in Phnom Penh.[1] When the Khmer Rouge took over Phnom Penh, in April 1975, they took over the radio station. The Khmer Rouge regime made propaganda that they distributed on the radio and in work camps: songs, radio programming, and films. By 1978, their programming included vicious attacks of the enemy Vietnamese regime, mixed in with incendiary reports of Vietnamese people and culture.[2]

During the Khmer Rouge regime, the United States, too, continued to wage war through media. The Voice of America (VOA) broadcasted into the country during the regime; some refugees reported listening to it in work camps secretly. Dith Pran reported to Sydney Schanberg of the *New York Times* after he migrated to the United States as a refugee that his commune chief had a radio. Schanberg reports that "sometimes at night—maybe once a week—Pran and four or five trusted friends would gather around it with him, and surreptitiously listen to the Voice of America."[3] During the war, VOA officials had no way to know if people were listening to their broadcasts within the country. They only heard that foreign-radio listening was a capital offense and that people did not have radios, or couldn't get batteries for them if they did have them.[4]

I am deliberately omitting a detailed history of the American bombing, the civil war, or the Khmer Rouge regime, and the ways that media played a role in these events and periods. In this omission, I seek to adjust an imbalance in postcolonial histories of Cambodia. We already have a number of histories and cultural products of Cambodia that focus exclusively on the wartime period, including Rithy Panh's work and much Western scholarship.[5] Though an account of the Khmer Rouge era is essential to understanding the history and contemporary conditions of Cambodia, other critical moments in Cambodian postcolonial history are sometimes overlooked. I will address this imbalance and the politics of memory in the second part of this book, particularly when I address another one of Rithy Panh's projects, the Bophana Center and its "Khmer Rouge learning app" in the final chapter.

Throughout my research, many of my Cambodian participants and friends shared with me unspeakably painful stories from their lives, including stories about the deaths of their families and violence they experienced during the Khmer Rouge period and the later war years. I have also interviewed and substantially interacted with admitted perpetrators of violence and people who continue to hold views of racial and ethnic hatred. These stories came out through living and working in Cambodia, often over the course of multiple interviews or experiences with participants, and were brought up by participants. As a general rule, I didn't directly ask interview subjects about their Khmer Rouge–related experiences or losses unless they naturally came up in conversation. Some of these stories are captured briefly in this book, while most are left out.

Listening to these stories has been a profound personal development experience for me; I have tried to become more skilled in the art of bearing witness to others' trauma. Active listening, for me, was important both within and outside the academic project; outside of the research, I was moved by the deeply human experience of connection that I felt in many interviews and other moments of research. I have tried to be an active audience to stories, as I have come to believe testimonies themselves can sometimes provide some catharsis or respite to trauma sufferers. Testimonies allow survivors to articulate their traumas, helping them find meaning in their incomprehensible experiences and feel a sense of agency.[6] The Transcultural Psychosocial Organization (TPO), the leading NGO in mental health care in Cambodia, follows this approach. They use testimonial-based therapy as their primary modality for easing symptoms of *baskbat*, the culturally specific trauma-induced condition of psychological distress in Khmer Rouge survivors.[7] I have also needed to learn strategies as a researcher and friend to not hold onto others' trauma (and become vicariously traumatized). This set of strategies I felt was useful for listening to, acknowledging, and working with my own and my participants' emotional experiences during research, including interviews and more casual encounters.

Ultimately, many participants shared experiences of trauma with me and these stories took varied forms. Throughout the course of the project, I began to categorize different ways people talked to me about their personal histories, how their own lives intersected with the Cambodian war period, and how they interpreted their pasts to relate to contemporary Cambodian politics. The history of trauma in Cambodia has been discussed and written about extensively, and some say the country is stereotyped for its violent and tragic history. Sometimes as a reaction to this, some participants (particularly younger ones) made a point not to foreground national trauma with me, and several participants explicitly told me their reasoning for doing this. Some older people, in recounting their autobiographies, would skip over the 1970s decade entirely. Some participants acknowledged familial and national trauma as an accepted given that they have always had to work their lives around. Others told me that they have had limited opportunity to talk about their traumatic experiences and so my attentive presence as audience amounted to something important. Some told me detailed, troubling, and gruesome accounts. Many participants were more interested in talking about what has happened since the Khmer Rouge, highlighting the strength of their communities and creative efforts, and how far the country has come. Others lament social problems in Cambodia today as products of its history. As I explore through this project, historical memory of violence comes in a number of forms and can be captured in either oral histories or historical artifacts, and it is often uncertain and sometimes repetitive.

I do not (and cannot) make claims to a truth of history that stands outside what was told to me by my participants or represented in their media creations. We must take care when presenting painful and traumatic imagery not to exploit the "pain of others" (Susan Sontag's phrase). There is a way in which trauma study can be voyeuristic or grounded in unacknowledgeable fascination or fantasies about victimhood.[8] The critic can also cast their analytical eye as "finer" than those suffering, as the sole person who can align the truth of the traumatic event with the representation of the trauma.[9] Suffering at a distance is routinely appropriated and commodified in popular culture (particularly American popular culture). It is important to avoid essentializing, naturalizing, or sentimentalizing suffering in this way. Images of suffering are appropriated to appeal emotionally and morally both to global audiences and to local populations in film and in the mass media. As Arthur and Joan Kleinman argue, "the cultural capital of trauma victims—their wounds, their scars, their tragedy—is appropriated by the same popular codes through which physical and sexual violence are commodified" in other forms.[10] Sometimes oversaturation of atrocity images also creates "atrocity fatigue" and immobilizes viewers.[11]

Media creators too struggle with the ethics of traumatic representation. Rithy Panh, who himself lived through the Khmer Rouge, struggled with the ethics of presenting suffering of others in his film *S-21* (where he brought together victims from Tuol Sleng and former prison guards and torturers). He said about the film, "The idea of putting victims and executioners together is very seductive, but it's also very tricky. You don't want to be a voyeur. You have to develop a kind of ethic of the image."[12] Other scholars analyzing traumatic media claim their work understanding and describing violent media can have positive outcomes and even mobilize political action. Caswell, who analyzes Tuol Sleng mug shots, argues that viewing the photos, when they are properly contextualized, can be a form of "co-witnessing." Publishing and digitizing the photos, and deploying them as legal evidence, can be the "highest form of respect."[13]

Bringing all of these experiences together, I decided to tell stories primarily of moving forward; this gives me a way to acknowledge trauma without dwelling on it or essentializing my participants for the suffering that they have often experienced. I hope these stories instead give us space to focus on strength amid pain. The focus of this project is therefore to discern the role of media infrastructures and their reconstruction in processes of post-conflict healing. So, rather than detail the ways that media were used as active agents of war during the most violent periods, I instead focus on the ways in which media infrastructures were reconstructed and the ways in which media were used in processes of rebuilding after the fall of the Khmer Rouge in 1979. I move toward this project in the next chapter, addressing the media context of the immediate post–Khmer Rouge context, beginning in 1979.

For readers who want more on the Khmer Rouge period, I suggest as a starting point Rithy Panh's 2014 documentary *The Missing Picture*, a compelling and personal narration of Democratic Kampuchea and its relation to contemporary media. This film combines archival footage of Khmer Rouge propaganda, narration, and clay figures to recreate the image of Pol Pot–era Cambodia. *The Missing Picture* demonstrates that materially rebuilding the "missing picture" of the Khmer Rouge period is fundamental to contemporary Khmer creation and future-building. For Panh, the "missing picture" is a visualization of what life was like for common people during the Democratic Kampuchea period.

The next chapter transitions to the period directly following the Khmer Rouge: the People's Republic of Kampucha (PRK) (1979–1991). During the bulk of the PRK period, the PRK government had a sometimes tense but powerful alliance with Vietnam; it also had strong ties to the other socialist states of the Soviet Union. Hun Sen was involved from the beginning of the regime and became prime minister in 1985. Though the Cambodian economy was officially collective and state-driven, state policy on collective farming shifted and loosened throughout the decade, and many communities slowly moved back into informal traditional family farming by 1990.[14] The economic tumult and postwar conditions led to instability, and the majority of Cambodians were impoverished during this decade.

Throughout the 1980s, in addition to the PRK, there were three other competing Cambodian political parties: the Royalist United National Front for an Independent, Neutral, Peaceful, and Co-operative Cambodia (FUNCINPEC, headed by Sihanouk's son Prince Ranariddh), the Khmer People's National Liberation Front (KPNLF, a right-wing, non-Communist, pro-Western party led by Son Sann, the prime minster from 1967 to 1968), and the Khmer Rouge. FUNCINPEC, the KPNLF, and the Khmer Rouge formed the Coalition Government of Democratic Kampuchea (CGDK) in 1982 despite ideological differences in order to stop what they called "the occupation of the Vietnamese" and had bases in Thai refugee camps along the Cambodian border. Under Cold War conditions (and anticommunist, anti-Vietnamese sentiment), Western powers supported the coalition and gave them the Cambodian seat in the UN, but the PRK ruled Cambodia.

2
MEDIA CULTURES OF THE PEOPLE'S REPUBLIC OF KAMPUCHEA (1979–1991)

In 2018, in collaboration with Sa Sa Art Projects, I ran a participatory art event in which we played a thirty-eight-minute *ayai roeurng* (comedy story), "Preparing for Cambodian New Year," performed and recorded in 1984.[1] Radio artifacts from this time period are rare, and I had collected and digitized the file at the National Radio archive, with the help of a longtime radio worker named Bou Vannarith, whom I will introduce later in the chapter. This genre of radio program mixes modern and traditional music with a comedy narrative and has been popular in Cambodia since the early postcolonial period. As those who lived through this period told me, in the 1980s, artists all sat in a room together at the Phnom Penh radio station and improvised these stories and songs. Our collective listening was a kind of a mirroring experience; as we were sitting together and listening, they were once sitting together and recording. We also mimicked the practice of collective listening common to many radio listeners in the 1980s and sat around the gallery main room on mats. We even listened on a device that resembled a 1980s radio but had a USB port so we could listen to the digital file. Before listening, we asked participants to consider questions around history and memory, materiality and degradation, and the experience of collective listening while the track played.

The program starts with a female radio announcer offering congratulations on the anniversary of the celebration of freedom from American imperialism on April 17, 1975, the day the Khmer Rouge took Phnom Penh. She suggests that listeners prepare for the Khmer New Year with extreme caution due to the risk of a Khmer Rouge attack. Interspersed with traditional *chapei* (traditional stringed instrument) songs, the core of the story is the tale of Yoo Sat, the village drunk who worries about his son who is "in

Figure 2.1
"Listening from the Archive." Sa Sa Art Projects, February 2018

the forest," the implication being that he is a Khmer Rouge soldier. Other details about life in 1984 are woven into the tale. Solidarity groups help a woman whose husband is in the military repair her home. The state youth military group helps to clean the village to prepare for the holiday. A guard watches the village at night to protect against a Khmer Rouge raid. At the end, Yoo Sat's son returns from the forest and decides to stay with his father in the village.

After we finished listening to the program, I explained to the audience where the artifact came from and showed images of the room of the National Radio archive where I found it. I showed images of the radio reel's cover, the materiality of the original radio reel, and the video of the process of digitization at the archive. We discussed the way the radio program had been made and the quality of the production, and how the production impacted the experience of listening. Some of the sound was stretched and slowed down because of the process of digitization.

Then the audience, my fellow organizers, and I reflected on the experience together. First, we talked about the experience of listening. One participant said:

> It was kind of cool. We always hear stories from our parents about how poor they were, how there was only one radio in an entire village. So it feels like I'm going back in time in this whole experience.

We also discussed the content of the story. Many participants said that the political messaging of the story was striking, and were surprised to see how explicitly the state used radio stories like this as a propaganda tool. These political messages were also jarring. For instance, some were surprised to see that in 1984, April 1975 was still celebrated as liberation from imperialism rather than a descent into a genocidal regime. We talked about how the story presented how difficult the war continued to be during 1980s. Another participant said:

> It wasn't really about preparing for the New Year—it was about being aware. It was about being careful and taking care of your family. It was about family members who went to the forest and disappeared. . . . It is . . . talking about the time that the Vietnamese government gathered people—young people—to be in the army and fight the Khmer Rouge. People lost their relatives and they lost their loved ones during the war. This isn't a comedy—this is a true story, and a sad one.

"Preparing for the New Year" is an unusual artifact from a moment in Cambodia's media history that is largely overlooked in contemporary Cambodia. Our participants' reactions point to the generalized confusion and lack of insight that many young people in Cambodia express about this time. This chapter recounts the reconstruction of media infrastructure during the PRK period (1979–1991) in the city centers of Phnom Penh and Battambang after the fall of the Khmer Rouge on January 7, 1979. The PRK can be understood as a liminal period, as it fell between the genocidal regime but before the end of the war. At the beginning of the PRK period, national infrastructures were shattered—including the media infrastructures of film and radio.[2] The chapter describes the process of reconstructing these infrastructures, focusing on radio in the first section of the chapter and on film in the second section.[3]

Media workers during this time fixed radios, cinemas, and transmitters and rebuilt media distribution infrastructures. They constructed a system of loudspeakers to play radio publicly, cleaned out cinemas (and developed material hacks like homemade air-conditioning), assembled mobile cinema teams (which ran even through active war zones), live-dubbed foreign-language films, and developed semi-legal home movie businesses. All of these activities constitute what I call *infrastructural restitution*, the creative reconstruction of media infrastructures.

Their work of media reconstruction included clear examples of Star's conception of infrastructural work, often-overlooked creative work that contributed to the country functioning in normal ways again.[4] We can see this creativity in resourceful material hacks used to repair broken infrastructures "brick by not-so-metaphorical brick" at a time.[5] This work was also highly emotional. Media reconstruction played a role in emotional recovery during the tumult of this immediate post–Khmer Rouge period. The emotional recovery inherent to the work is evident in certain artifacts of media content. For example, the national radio programs such as the Motherland Appeal (Neaty Somleng Ompeaveneav Robos Meataphum) were made for the purpose of national reconciliation, forgiveness, healing, and family reunification; certain films (e.g., *Rodau Pka Tnaout* [*The Season of the Palm Flowers*], 1980) helped both the media creators and their audiences to recover, heal, and move forward from the Khmer Rouge. We also see examples of these workers' bravery, as they endured violent conditions in order to make space for entertainment and communication in a volatile world.

Analyzing these forms of infrastructural restitution also makes clear the power dynamics inherent to the media infrastructures of this time. Though the recovery work can be understood on a personal, affective level, it can also be seen as a nationalist state project as the PRK party fought to be recognized as the legitimate government of Cambodia and enacted control through censorship policies. Working in part from Hecht's concept of technopolitics, this history makes clear the geopolitics of technology during the PRK period and the role of non-Western foreign powers (particularly Vietnam and the Soviet Union) in the development of a socialist media in Cambodia in the PRK.[6] These power dynamics are not only at the scale of the nation but also at the scale of the individual worker, as media reconstruction happened within a context of many other inequalities and violence. Many Cambodians—including some of these media workers—lived in poverty and were enduring frequent but often unpredictable interparty violence. Others had stable and relatively safe jobs within the state.

This chapter is based on archival material from the National Archives of Cambodia, the Foreign Broadcast Information Service (FBIS) archive, the

Figure 2.2
Bou Vannarith at the National Radio, September 2017

Bophana Center archive (including PRK-era films), secondary historical material related to the 1980s period, and oral histories.[7] The oral histories of four primary actors—Ka Toy, a projectionist and repairman from Battambang; Bou Vannarith, a former employee of the National Radio; Loak (Mister) Kamala, who ran a video screening company in Battambang; and Mao Ayuth, a secretary of state at the Ministry of Information and a famous film director—make up the heart of this chapter. Like all oral histories, these interviews should be understood as perspectives on a politically complex history, filtered through the memories of men living in a politically complex present. Their status as state workers, their identities as men, and their varying amounts of political and economic resources all are relevant to understanding their interpretations of the past.

1 PRK RADIO INFRASTRUCTURE

In an interview in January 2018, Bou Vannarith, a recently retired radio employee, told me, "I started working for the National Radio in July 1979, a few months after the liberation from the Khmer Rouge regime."

Recruitment for the National Radio after the Khmer Rouge was a challenge. "There were a few rare—only a few—technical people who survived the Khmer Rouge regime." He continued,

> There was an announcement—it was spread all around the provinces—looking for technical people for the radio. They also had a small, small car—they put a speaker on the top and they made an announcement to look for technical men and women who could run the radio station. They asked to please come here to the National Radio to help. I heard the announcement on my way back to Phnom Penh and that's when I decided to come to the radio for the first time. The person I first met at the radio liked the sound of my voice. I can read perfectly . . . and have the accent like Cambodian leaders.

At first, he was given the position of news announcer. Vannarith continued,

> At that time, in 1979, there was just one radio—the National Radio. Before the liberation, the station was run by the Khmer Rouge. After the 7th of January, the Khmer Rouge dispersed. All the soldiers left. However, the radio—the technical equipment—was very old, and the network was very old. We had to repair it all. . . . After the liberation, we took this place [the current National Radio building where we are conducting the interview] as the administrative center. . . . We used the building [closer to Wat Phnom a few blocks north] for broadcasting . . . it was only 100 meters from here.[8]

On February 5, 1979, the Voice of the Cambodian People run by the newly established PRK government started broadcasting on the same frequency and from the same transmitters and studios that the Pol Pot regime had been using in Phnom Penh, less than a month after the fall of the Khmer Rouge in the city on January 7, 1979.[9] The nascent PRK government was pouring resources into the radio station, which Un Dara directed. Un Dara saw rebuilding the radio as a key part of reconstructing the nation. As a part of an announcement of a technical radio training, he stated on September 23, 1979: "After liberating the country on 7 January 1979, we were faced with the task of rebuilding the country in all fields and *particularly in the field of radio technology*. Under the Pol Pot–Ieng Sary regime, much of the equipment had been destroyed and many technicians massacred. Now we open this course in order to teach our trainees radio and television technology."[10]

The "old people" at the radio taught the "new people" like Vannarith to write the news and run programs. Vannarith explained that "in the live broadcast, at that time, we talked about all sorts of things. The boat festival, the boat racing, the national Independence Day, the anniversary of the 7th of January." Back in 1979, Vannarith explained, "we broadcasted eight or twelve hours a day. In the morning, we started at 5:30 and we stopped broadcasting at 9:00 am. Then we started again at lunch from 11 am until 2 pm. Then we broadcasted again at nighttime from 6 pm until 10 pm."

The old people also taught him and other new people to run the station. Just two years after starting to read the news, Vannarith was promoted as a deputy of the office of programming. "At first, we only focused on the news." Once they had a solid news broadcast, he explained, "the second thing, we focused on the music. . . . Step by step, we rebuilt the radio. We created a lot of kinds of programs in the end."[11]

The PRK government knew that people of various nationalities were listening to their radio station, including those who did recognize their government, and used programs to emphasize their legitimacy. In July 1979, the radio began some broadcasts in English, French, and Thai.[12] Un Dara greeted foreign listeners and told them that "following the collapse of the fascist dictatorial regime of the Pol Pot–Ieng Sary clique, the People's Republic of Kampuchea came into existence. Under the leadership of the People's Revolutionary Council—the sole authentic and legal representative of Kampuchea—the Kampuchean people are making all-out efforts to rehabilitate and rebuild the country."[13]

The Vietnamese government supported the radio station heavily, including through technical trainings. In early 1979 and again in September 1979, the PRK government, through support of the Vietnamese, offered a "radio technology course" to hire and train more radio technicians.[14] The head of the Vietnamese advisors to the Cambodian radio, Nguyen Vu Cap, said, "Despite our shortages, we have established our radio technical ranks, including radio-broadcasting acoustics and electronics. We will make every effort to teach and turn Kampuchean youths into experts in radio and television technology."[15] The PRK Minister of Information, Press and Culture, Keo Chanda, told the trainees "to pay great attention to learning radio and television technology and become experts in this field to help build our Fatherland."[16] The first cohort of this training program ended in 1979 with twenty-two graduates, and the second ended in April 11, 1981 with 110 graduates.[17]

In the years directly following the Khmer Rouge, the PRK and Vietnamese governments also invested heavily in rebuilding radio cables and repairing transmitters. In September 1979, the Vietnamese government helped the Cambodian government install a new radio station in Takhmau, 10 kilometers south of Phnom Penh, with a network of 10 kilometers of wire and fifteen loudspeakers.[18] The party opened a radio receiving station in Battambang in January 1980 and in Prey Veng in February 1980.[19] The central government reported in 1983 that it consistently increased the distance of the network cables and repaired equipment in the first five years after the end of the Khmer Rouge, from 52 kilometers of network cable to 88.5 kilometers.[20]

The party fought hard to maintain radio infrastructure, in part so that they would be the dominant voice in the radio audible in Cambodia. As in the early postcolonial period, opposition parties weaponized radio range across the border. Though the state radio was the only station clearly audible in Phnom Penh, different parties competed for radio waves in border zones; they developed radio stations from their area of control, so these could be heard depending on the Cambodian region and proximity to the border.

In 1981, the Chinese government gave Khmer Rouge leadership, then based in Yunnan province, a "massive" radio transmitter. The Khmer Rouge was able to broadcast into northern Cambodia and often reported about battles and gains in territory.[21] By 1983, the Khmer Rouge ran two stations: the "Voice of Democratic Kampuchea" and the "Voice of the National Army of Democratic Kampuchea."[22] Later they moved their radio operation to the Pailin area, which remained clandestine and broadcast twice daily until 1996.[23]

In October 1982, the KPNLF, led by Son Sann, started a radio station in Thai refugee camps.[24] After some difficulty establishing a collaboration, the Sihanouk-aligned FUNCIPEC party joined forces with the KPNLF to start a new, joint radio station called "The Voice of the Khmer People" (VOKP), broadcasting from two mobile transmitters on the border in December 1983.[25] In an article in the *Bangkok Post*, representatives from the parties said they needed to have a mobile transmitter because its indeterminacy of location protected against attack from the ruling government and Vietnamese troops, even though it led to less powerful transmissions.[26] The article reported that the VOKP could broadcast 300 kilometers into Cambodia from the Thai border, or as far as Kampong Chnnang.[27] In January 1984, a "ranking KPNLF official," quoted in the *Nation Review*, asked diasporic

Cambodian technicians to work on the "extension of the broadcast range of a clandestine and mobile radio station . . . describing the radio transmission as a powerful tool against the Vietnamese occupation of Kampuchea."[28] Elsewhere, officials from these parties stated that the purpose of the radio was to "assist the nationalist Khmer resistance fighters" and "preserve our heritage."[29] The strength of the transmission was a tool of political persuasion. Like the PRK party, the KPNLF and FUNCIPEC emphasized the importance of radio in fighting for their own legitimacy.

Each party's radio station had different perspectives on the past, reflected in their different content. The VOKP strategically played songs from the Sangkum Reastr Niyum era to remind people of the old regime.[30] These "old regime songs" from before the Khmer Rouge (particularly songs from the Sangkum Reastr Niyum) were not allowed during the early period of Vietnamese occupation, due to their association with capitalism and imperialism. Vannarith said, however, that "five years, six years, seven years after the liberation, we started to mix the songs. . . . [We could start to play] both the old regime songs and the new songs."

Once the PRK radio started playing songs as well as programs in the early 1980s, they started by only playing songs from what they called New Serey Recording, or songs from the Vietnamese occupation era. Vannarith explains that in the 1980s "all the artists" in Phnom Penh worked for the radio station. Radio artists often recorded original songs in the studio as part of *ayai roeurng* (radio comedy) and *Ikhoan niyeay* (spoken theater). Some of these works are beautiful and original works of commemoration. These are rarely played today because they were lost or because they can disorient contemporary listeners with their inclusion of communist language.[31]

This music on the PRK radio was used as a vehicle for national reconciliation. Vannarith and his wife Ros Kandavy ran the Motherland Appeal radio program on the National Radio, which called Khmer Rouge soldiers back to Phnom Penh in the 1980s. Ros Kandavy interviewed former Khmer Rouge soldiers either in their new, resettled villages or in the studio, and explained the terms of the government's forgiveness to all defectors. They also occasionally used the program for reuniting relatives or returning belongings to people who owned them before the Khmer Rouge.[32] The opening song for the program, called "Sronos Srok," is a sad traditional Cambodian flute song. It appealed to moods of prewar Cambodia, though is more traditional than the modern music of the 1960s and early 1970s that was played on

VOKP radio. The program focused on leniency and forgiveness for former members of the Khmer Rouge. It played twice per day from 1981 until 1989. Family members or loved ones could put announcements in the program looking for those who might still be in hiding.

The Motherland Appeal program was far from politically neutral for the Khmer Rouge, KPNLF, and FUNCIPEC parties. The Voice of the Khmer People broadcast said about the program:

> In its daily broadcast, Vietnam orders the puppet regime's radio to organize a program entitled: The [Motherland's] Appeal. This appeal program is accompanied by melodious and nostalgic Cambodian music. The announcers also use sweet words which should make listeners feel homesick for their beloved fatherland. However, the aim of this appeal program is to rally Cambodian resistance fighters to serve Vietnam. . . . The Cambodian music accompanying this appeal program has made us nostalgic for our Cambodian fatherland, against which the Vietnamese enemy is committing aggression and is trying to gradually wipe out Cambodian culture until it completely destroys everything Cambodian.[33]

As in the Sangkum Reastr Niyum period, the radio had again become one of the key tools from which Cambodian political parties espoused their messages to Cambodian people. But a question remains about how many Cambodians were actually listening to the radio during this period. Many participants told me that it was uncommon for individual families to own radios during this time because of widespread poverty. In Phnom Penh, many Cambodians primarily had access to goods from the Eastern Bloc countries and there were a limited number of radios and TVs for sale. There was an active black market for buying media equipment on the border with Thailand, particularly adjacent to refugee camps.[34] These goods, however, were expensive when available. I often heard the anecdote that people used gold from before the Khmer Rouge that they had buried and later dug up to purchase commercial black market media products.[35]

Nevertheless, Vannarith told me that he expects that through most of the 1980s period that "70–80 percent of the Cambodian people could listen to our national radio. There were not personal radios. [Cambodian people] had small, battery-operated radios in the villages. There were not enough—not everyone had their own radio—so people listened together. So, for example, you are here—you have a house—many people come to your house to listen to your radio in the village, in the province."[36] Some people also had radios from the old regime or radios donated from the

USSR. These radios were big and required large batteries. An interview participant in Battambang told me that there were challenges getting the radios to run since most radios needed a lot of batteries—he claimed more than twelve batteries for one radio, which would only last ten days—and batteries were expensive. Each day for the ten days that the batteries lasted, they would try to recharge them by putting them in the sun and then using the radio only at night. Vannarith said that people would power radios using their bicycles, which could make enough electricity to charge their radio batteries.

In order to increase the number of radio listeners, the government also installed networks of speakers in city centers and some villages. When the National Radio started to broadcast in 1979, Varrarith explained, "we could all listen to the network speaker—the authorities wanted the people to listen." The Phnom Penh authorities put speakers on poles in main locations and connected wires to a series of other speakers. There were speakers installed around Wat Phnom (a major landmark near the National Radio) and around Psar Tmai (the central market) for at least five years. Vannarith explained, "It was so funny—you could sleep in this room [referring to the radio station], but you could listen to the radio from the Wat Phnom speakers." Government statistics reported that there were forty-six installed public loudspeakers in Phnom Penh in 1979, which increased to 112 in 1983.[37] Networked speakers were not limited to Phnom Penh but were installed through provincial capitals such as Battambang, with the assistance of the Vietnamese.[38] A participant in Battambang told me that the government would also play radio from speakers there so that people could gather at the riverside and night market and listen together.[39]

The reconstruction work done by media workers like Bou Vannarith made up a kind of infrastructural restitution—the creative repair of media infrastructure. Material concerns, such as the strength of transmitters, the installation of loudspeakers, and the development of new radio stations, were seen as key to successfully reconstructing the infrastructure. Technicians like Vannarith also had to train quickly as many former technicians had died during the Khmer Rouge period. Workers often demonstrated resourcefulness; they rebuilt skills through mentorship and technician training programs. Conditions of war and recent genocide also made this work at times highly emotional; the mournful element of the work is clear in frequent nostalgic and sorrowful radio content.

This large infrastructure project—the installation of loudspeakers and public broadcasting of radio—also demonstrates how important the radio station was to the PRK government as a tool to demonstrate its legitimacy and its ability to lead national reconciliation projects. The state had the authority to play its programs loudly in city centers in lieu of competing broadcasts. Hun Sen, who became prime minister in 1985, explained in a speech in 1988 his philosophy on the centrality of radio in PRK-era Cambodian national life: "Dear comrades and friends . . . Without news and dissemination of information, we cannot talk about the intellectual and spiritual growth of society. . . . I would like to take this opportunity to praise and highly value elder brothers and all the comrades, the first artists, for laying down the first brick for the VOKP broadcast. I would like to praise the precious and effective assistance of all the comrade Vietnamese specialists who made the first broadcast successful."[40]

This nationalist project continued to have transnational support, and the cooperation between the Cambodian government and Eastern Bloc countries in matters of radio continued through the 1980s. Delegations of Vietnamese radio experts came to Cambodia regularly to support the station and a protocol on cooperation was signed as late as 1990 between Vietnamese and Cambodian radio technicians and musicians.[41] The USSR and aligned countries also increased their support for information and propaganda, with Cambodian radio technicians traveling to Moscow in 1983 and Czechoslovakia in 1984.[42] Czechoslovakia donated broadcasting equipment in 1983.[43] The USSR declared memorandums of support for radio tasks from 1984 to 1986 and 1987 to 1989 and a Soviet radio delegation visited Cambodia in February 1990.[44]

2   PRK-ERA FILM INFRASTRUCTURE

In addition to radio, film was another important media infrastructure that the PRK government worked to reconstruct in the immediate post–Khmer Rouge period. The PRK's Department of Culture and Propaganda began a film program in 1979 under the direction of Mao Ayuth (who died in 2021) and Ieu Pannakar (who died in 2018). Ieu Pannakar was one of the first filmmakers that Sihanouk sent to France for study in the 1950s with Som Sam Al. Mao Ayuth had first worked with him from 1963 to 1965, when Pannakar taught screenwriting under Sihanouk's direction. After the

Figure 2.3
Mao Ayuth at his desk in the Ministry of Information in January 2019, age 75, holding the novel of his film *Beth Phnek Hek Troung* (Close My Eyes, Open My Heart)

course, Mao Ayuth then worked for the Télévision Royale Khmère (TVRK), founded by Sihanouk, from 1965 to 1970 (which he told me "never had a big audience"). He studied and worked in France from 1970 to 1974 and made his first feature film, *Beth Phnek Hek Troung* (Close My Eyes, Open My Heart), in the Swiss Alps. Only one print of the film was made, which was shuttled from cinema to cinema in 1975, just before the fall of Phnom Penh. Though the film was lost during the Khmer Rouge period, it was so popular that Ayuth wrote a small novel based on the story, of which he has a remaining copy (pictured in Ayuth's hand in figure 2.3). Ayuth survived the Khmer Rouge period in Cambodia by pretending to be a wedding photographer and passing a lie detector test. He lived through the period as a fisherman and laborer, when most of his former colleagues from film and TV died.[45]

Mao Ayuth recounted his memories of rebuilding a film program in a January 2019 interview:

> I was in charge with Ieu Pannakar of the cinema program after 1979. Many people had moved abroad so I started mobilizing the people who liked film

[to gather a team]. There were cinemas at that time but only one or two in Phnom Penh running. We had to clean the cinemas after the Pol Pot regime. We had to check if the seats were broken and we had to get rid of rats. Some people were also living in there. The Mean Teap cinema was the first, which they changed to call Brajeajun Cinema [People's Cinema]. Now it is Tous Les Jours bakery. I used to be the director of that cinema. . . . There was no aircon at that time. We announced to people that there was aircon and they used the fan with water to make like an aircon. . . . In the cinema this kind of aircon was available on the balcony but the ground floor, they got sweaty [*laughs*].

Other cinemas slowly started to open around this same time. For instance, in January 1980, the Soriya movie theater opened again with a screening of a documentary that denounced the Pol Pot–Ieng Sery regime.[46]

The PRK government often hosted elaborate, state-sponsored film screenings in Phnom Penh. On Sunday August 30, 1981, on the anniversary of "the 36th National Day of Vietnam," a film week opened in Phnom Penh with the presence of the Cambodian Minster of Information (Chheng Phon) and the Vietnamese ambassador (Ngo Dien).[47] One interview participant told me that there was a Vietnamese hospital on Street 172 between 63 and 51 that played Vietnamese films every two or three months. Nearby on Street 174 near street 63 there was an old jail that played Soviet films at the same frequency. Film festivals featuring films from various Soviet countries occurred regularly. A government report from 1983 showed that film showings increased from 0 in 1979 to 1,840 by 1983, with audiences increasing from 0 to over 2 million in 1983.[48]

Ayuth continued to explain the development of the film program:

The government wanted people to see film because there was no film during the Pol Pot regime. I was the one who got to choose which films to play from Vietnam and Cuba; these countries normally invited me every year to see their films. I tried to choose the movies that were best contextualized to Cambodia such as *Ali Babba* and *Pka See Manou* (Flower Eats Human) . . . there was a big giant in that one and I chose the giant for Cambodians [because Cambodian audiences are known to like special effects]. There was a movie with snow, but I didn't choose it because no Khmer people would watch it because they don't know the snow. . . . Since they were in Russian or other languages, we had to translate them into Khmer using voice actors.

When we started showing films, we would show one film for a month and people would just line up . . . we wanted to teach people how to line up to see movies. While waiting, some people might lose their belongings, like their bracelet or necklace. Some people would sell their property in order to see movies at that time. The tickets were very cheap, five cah, one riel . . . and right after it finished we sold *bar* [rice snacks] to make some more money. We needed to help each other to find the money at that time. The money bills were old.

Some remember the films from this period with ambivalence. Reach Sambath, chief of the public affairs section of the Khmer Rouge Tribunal, said in a published interview, "After the fall of the Khmer Rouge regime, when I was in grade 5, most of the films that were screened in Cambodia came from socialist countries, like the Soviet Union and Czechoslovakia. Most people were still afraid of going to the cinema, because there might be bomb attacks by minor Khmer Rouge soldiers that still tried to create chaos in the country."[49]

Mao Ayuth said that people most preferred to watch Khmer films instead of foreign films. There were not many Khmer films made during this period, and almost everyone reports in contemporary Cambodia that these were not as high quality as those from the 1960s or early 1970s "golden age." Mao Ayuth and Ieu Pannakar themselves made many of the Cambodian films from this period that do exist. *Rodau Pka Tnaout* (*The Season of the Palm Flowers*) is a black and white twenty-four-minute film made in 1980 by Mao Ayuth and Lim Kvang Ngoc.[50] The film depicts the agricultural process of making palm sugar, including peasants collecting palm sap in a bamboo shell and boiling the sugar. The filmmakers show a sugar factory in Kampong Speu, which was repaired with funding from Czechoslovakia. The film flips to photos of what they call the "Pol Pot–Ieng Sary" time, which they depict hellishly and which they juxtapose with stills of the strong PRK soldier. The film returns to photos of happy-looking people eating palm fruit, pictures of the ocean in Kep, and kids meeting their friends then going to school. The film finishes with images of a still (relatively empty) Phnom Penh in 1980 and people riding around on bicycles and cyclos. The film cuts to the countryside, where people drink palm sugar, which they bring to the city to sell to city dwellers. The images of agriculture and children are clear symbols of rebirth, reconstruction, and nationalism after an intense period of violence.[51]

Mao Ayuth made his first fiction feature film since the Pol Pot period in 1988, called *Chet Chong Cham* (I Want to Remember) (1989), a story of survival set during the Republican period under Lon Nol (1970–1975), under the Khmer Rouge, and in the 1980s, told in flashbacks within flashbacks.[52] It was shot on an analog video camera with a $400 budget.[53] The lead actor, Kai Prosith, performed for free and became one of the most popular actors in the 1990s. Mao Ayuth said about the film, "Sometimes during the Pol Pot regime, the regime asked me to drive an oxcart. I didn't feel any nervousness at that time. . . . I just walked and observed the nature around me when I used the cart. I just thought about when the regime would collapse. I didn't know how it would collapse, if it would collapse on its own, or if another external party would come and help . . . because during that regime, there was no money, no school, so I was carefree at that time. I just watched a mountain, the tree wherever I would go. I decided if the regime would collapse, I would make a film about it." *Chet Chong Cham* became the film that he dreamed of making during the Khmer Rouge.[54]

Ka Toy (a nickname) is an electronics repair-person and former projectionist who lives in downtown Battambang city. His shop is next door to the old (now defunct) Battambang cinema in an old colonial townhouse in the center of the city. He moved into this building from his first repair shop, only about a kilometer from his current shop, in 1996. The bulk of his business today is repairing amplifiers. People from all the northwest provinces come to his shop when they need an amplifier repair. He also makes new amplifiers from spare parts that he collects, which he then puts on sale in his shop.

Ka Toy knows how to repair the amplifiers from his experience working as a projectionist in the 1980s. At that time, he installed all kinds of equipment, repaired equipment, and taught himself how to understand how circuits work. From that job, he was able to teach himself how to fix and build more complicated amplifiers. He started his first repair shop in 1992, once his work as a projectionist dried up during the UNTAC period.

Before the Khmer Rouge, his father worked as a guard in the Battambang prison. When the Khmer Rouge took Battambang, soldiers evacuated him from his home behind the former jail. During the Pol Pot period, he was forced to work at Phnom Sompeu.[55] He worked on a mobile unit for youth and helped build two dams; his parents and grandparents all died. He told me that he has only a few relatives now. "All other cousins are gone. I just have two siblings who survived. . . . I would have left the country after the Khmer Rouge but Son San, Sihanouk, the Chinese, they are all the same." After the Khmer Rouge, at the age of seventeen, he moved back to his house

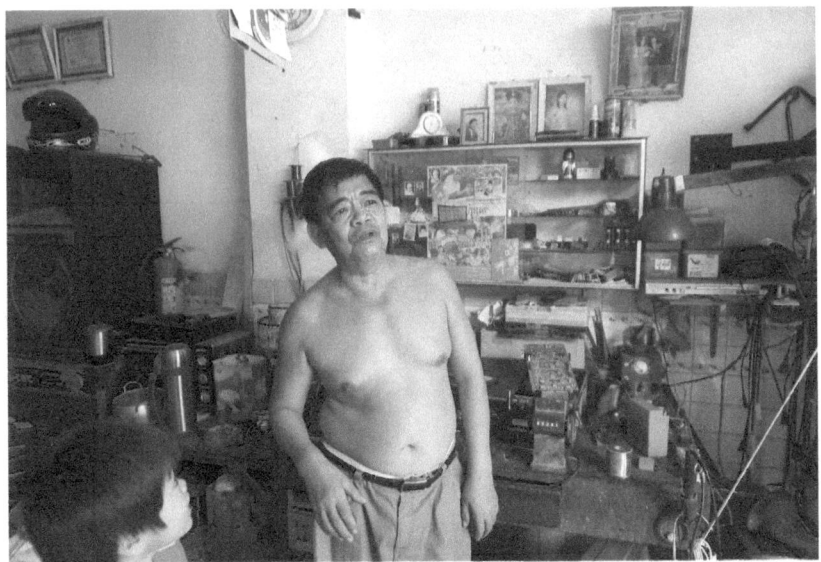

Figure 2.4
Ka Toy in his shop with his grandson in August 2017

in Battambang. When I asked him how he got the job as a projectionist, he shrugged. He had received a diploma from the Lon Nol period and he "was an adult." He had a relative who was a projectionist before the Khmer Rouge period. He recalled, "There weren't too many people around who knew anything about this sector." He was asked to go to the Soviet Union but he declined, saying that he would rather stay in Battambang. For the next ten years he worked for the state as a projectionist.

Throughout the 1980s, Ka Toy had different jobs at the cinemas around Battambang, which were all built from the French colonial era through the 1960s. He worked at a cinema referred to as Hab B through the Sangkum Reastr Niyum period, though during the 1980s it was officially called the "7th January cinema." By the time we spoke in 2017, the former theater had been converted to a club. He also worked at the Battambang cinema from 1982 to 1983, which, by 2017, had been abandoned. In 1984, he started projecting films at the Hap Chouen cinema, which, in 2017, had become the Battambang town restaurant. He was the projectionist there for seven years. All these cinemas were within a few blocks of each other and were very popular. He also repaired equipment. He said that, though he

Figure 2.5
A group of mobile projectionists including Ka Toy (top row, center), early 2000s. Source: Ka Toy's collection

was aware of the training for media specialists in Phnom Penh run by the Vietnamese, he chose not to go and instead learned on his own.

Through the 1980s, Ka Toy also had a lot of work as a mobile projectionist. He traveled all through the western provinces and showed state-sponsored films to large crowds. Ka Toy recalls that they traveled "by the land, by the wats, by the villages, in the army bases." At that time, they played films from the Soviet Bloc—films from Russia, Vietnam, Czechoslovakia, and other countries. He remembers that sometimes the Eastern European films featured movies about World War II. People liked these films a lot and the showings were always very crowded. Since the movies were in other languages (primarily Vietnamese or Russian), the group would travel with live translators who would act out the films in Khmer. They also traveled with guards. The team traveled in military trucks or horse carts. He told me, "We would bring the machines on the horse carts and some of the team would travel by bicycle. We used a Soviet machine. We needed a generator, and we would use the old ones from the Khmer Rouge time or before, provided by the state."

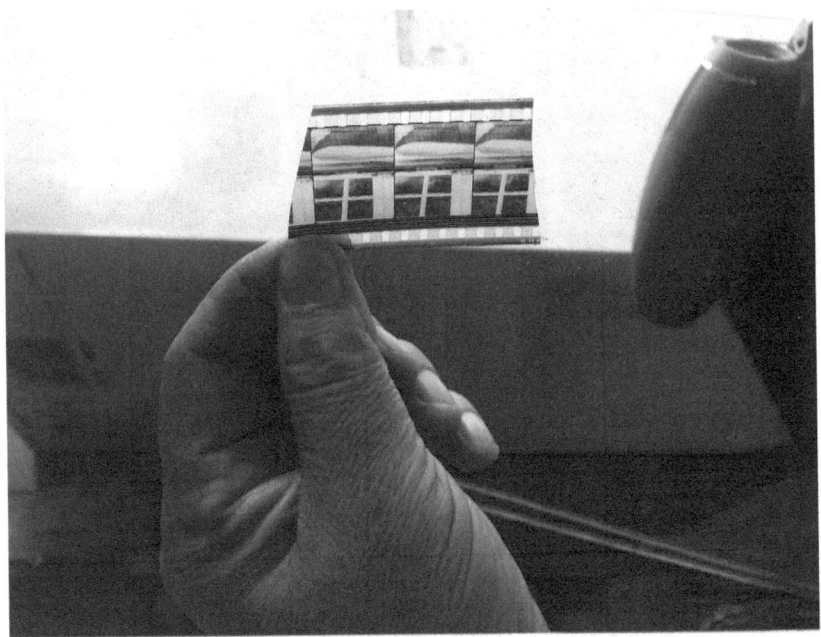

Figure 2.6
Film showing the sound strip, which they could turn on and off for the actors

Mao Ayuth, who directed these film screenings from Phnom Penh, told me that they were necessary for getting films to rural communities: "The mobile cinema was a good business at that time because people were poor and it was right after 1979. . . . The groups had to travel to rural areas nationwide to show the films. It was very difficult to travel in the provinces and we had to bring a generator. It was funny. Among the twenty teams, they had to bring a voice actor and sometimes one actor had to perform all the people, young people, a man, a woman. . . . At first the voice actors were recruited from the Ministry of Culture. After that, we empowered the provincial officer to choose the voice actor."

Ka Toy showed me a piece of film from this period (see figure 2.7). He explained: "This is a Soviet film and the Soviet sound is here [pointing to the top of the film]. We had a switch so that we could play the sound when there isn't a line. When there was a speaking line, we would push the switch—the sound would go silent so that a Khmer person could talk." The lines were read from a book of translations written in Phnom Penh. Ka Toy

Figure 2.7
Ka Toy showing me some of his old fragile film

remembered these live translators, laughed, and said that sometimes the translator could not do it well because the movie was too fast.

In the 1980s, there was still some fighting in these western provinces. Ka Toy said, "At the same time I was projecting, I was asked to shoot." He continued, "There was one movie screening at Wat Lawea when I was really scared. Usually they had a movie program and then dancing. The movie ended and they were about to start the dancing. The Khmer Rouge fired a B40 rocket launcher inside. It didn't explode and only the hand rifle was firing. They tried to find me, the movie screener, but I jumped into the river instead."[56] Ka Toy laughed again.

Fifteen days after he was married, in 1984, Ka Toy was asked to be part of a media mission to call the Khmer Rouge back to join the PRK government. He explained that the party called to Khmer Rouge soldiers on a loudspeaker: "Come, surrender, you will not be punished. We will forgive you." Ka Toy told me, "Yes, they came, but with machine guns—and they fired at us."

I asked him why movies were so important to him to face this kind of danger. He responded, "We worked for the state. In that time, working

Figure 2.8
Old film reels and equipment in Ka Toy's attic

for the movies, it was an easy job. Working for the government, you got enough money. It was a pretty easy life. I didn't want to be in the military. Otherwise they would arrest me and make me be a soldier." I asked why people would go to the movies if there was a chance of violence. He responded, "Because they didn't have anything to watch. In the cinema, sometimes they could screen one movie for the whole month, four times per day. People would still go to watch it."

After our first interview, we walked upstairs to the attic of the shop where he keeps his collection of old projection equipment and old film reels. None of the films there is a classic Khmer film—he has sold these throughout the years to collectors.[57] The films that remain are from the United States, Thailand, and India. They are too fragile to play now. "Right now I am just storing them," he said. After a pause, he shrugged and continued, "Maybe later I'll get rid of them." He has a 16-mm and 35-mm projector, both of which are beyond repair. I ask why he keeps the material. He said, "I keep them because I love them." He has great memories of projecting films and misses the way the film industry was. He keeps one of the projectors in his house so that he can show it to his children.

Ka Toy said the younger generation has very little interest in watching film. "Because of the rise of new modern technology, young people today like to watch movies on YouTube, on their phones." I asked him if he thinks it is important for them to see old films, and he sighed. "Oooh. I don't know. Everything is modernized now. If we want to attract the young generation, we have to develop movies to fit their interests . . . the young generation, they like the 4D, 3D, 9D films!"

Another avenue for film watching during the PRK period was home TV cinema businesses. A resident of Phnom Penh said in a published interview, "In 1985 or 1986, there were no Cambodian films. When I went to a cinema at that time, I could watch only Indian and a few Chinese and Vietnamese films. Cinema at that time was not like the cinema nowadays. It was often just in somebody's flat. We sat on wooden chairs like in school."[58] In March 2018, I met Loak Kamala, now retired from a long post at the Ministry of Agriculture, at his home in Battambang. I met him through a friend who now lives in Phnom Penh but grew up in Battambang, watching Kamala's TV in a home cinema business.

Kamala explained to me that in the early 1980s, before he had his own TV, he would watch TV at his neighbor's house. "I didn't like making my neighbor's house dirty." So around 1986 he bought his own TV to start his own business. "Luckily my mother had enough money—the TV was so expensive. It cost one *domleung*—maybe $1000 now. We bought the TV new at Battambang market and it was in bad quality color." He explained that "in that time, salaries were not high—you could get rice in place of salary." Kamala collected money from people who wanted to watch the TV at his house. People had to pay some coins (one or two coins, or about 100 riel) to watch TV in an evening. "Many people came to watch movies at my house every day . . . it was so crowded!" After a year or so of having the business in his house, he made a deal with the local pagoda. There wasn't enough room any more in his house, so he moved his TV to the pagoda and split the profit with the pagoda.

During this time, though, the business felt dangerous. Kamala said that people were always afraid and sometimes were downright uncomfortable watching the movies. Boys or even "all the people" would run away when soldiers came to try to recruit men and boys to be soldiers. Kamala explained, "Sometimes even Khmer Rouge soldiers—teenagers, boys— would come to watch. While they were watching, the police would come to catch the boys to be soldiers for the government. The boys would yell so loudly to make their mothers know the police caught them. The yell meant the police would make him be a soldier."

Kamala explained that sometimes rich people would just rent his TV to watch it in their village. Often these were rich people in the gemstone and diamond business on the border regions. They would watch TV "not just for business—it was for escaping; it was fun and a happy time." But, he explained, "they would have no security when they watched TV; they just made a loud sound if an opposition party were coming. If someone came and they felt afraid, they would run away." One time this happened and Kamala became injured as the crowd dispersed. His facial expression during our interview expressed how difficult and painful this memory was for him.

Despite safety concerns, Kamala and his crowds would watch foreign movies from India, China, and Hong Kong. While there were no stations on the TV, they could play videos, which they rented from the market as videocassettes. Each rental was 5 or 6 riel. These rental videos became available in Battambang around 1985, and more became available in 1987 and 1988 as markets slowly opened up to more commercial goods.

Kamala's family paid a photographer to take photos for their home cinema business. One photo shows his two boys in front of the TV and one shows a few people hanging around the house about to watch TV. At that time, Kamala said they didn't have many cameras. While we looked at the photos together, he told me that "by the late 1980s, as it got closer to UNTAC time, they had more cameras, and more in color. The better [cameras] came from Thailand."

Gottesman recounts a similar home cinema business in Phnom Penh that became a central point of tension among Phnom Penh politicians, and is captured in their internal meeting reports.[59] At the end of October 1985, Yen Sok Ieng purchased three videos for home viewing in Phnom Penh, what are called in these reports a "violent one, a romantic one and a pornographic one." He and his brother Yen Song Huot charged 15–30 riels per showing. In November, Phnom Penh authorities shut down the business and arrested the "two Yens" and the man (Sok Bun) who sold the videos to them. This was characteristic of the PRK censorship policies against playing Sangkum Reastr Niyum–era media. By the end of 1985, the police had seized 167 pornographic videos or videos from the "old society" of the Sangkum Reastr Niyum. The "two Yen case" and their videos, however, became the subject of "intense debate at the highest levels of the Party."[60] Ultimately, the case faded away, but it demonstrates how seemingly trivial questions of media were central to state workings and how the PRK saw information control as a key to political power.

The reconstruction work that people like Bou Vannarith, Ka Toy, Mao Ayuth, and Loak Kamala were doing is a kind of infrastructural restitution, or the creative reconstruction of media infrastructure. Part of this work was repairing or constructing the material aspects of the infrastructure—the transmitters, cables, and loudspeakers. Another part of the work involved building logistical chains for the performance of mobile film in rural parts of the country. As in the Sangkum Reastr Niyum period, the material constraints of the stuff of infrastructure (broadcast lines, transmission strength, networked speakers, access to villages for film screenings) played a large role in the way messages were received. Just as the PRK strove to build stronger transmitters and a system of receivers, the Khmer Rouge and KPNLF/FUNCIPEC worked to build stronger transmitters on their sides of the border.

Speeches from the time and the memories of those whom I interviewed demonstrate that rebuilding radio and film infrastructures was central to the state reconstruction of the country after the Khmer Rouge period. For the PRK government, radio, film, and TV were critical infrastructures of nationalism, reflecting legitimacy and educating the population about their political positions, while also providing entertainment and sentimental content. This chapter thus highlights how power structures can be present within processes of reconstruction.

The reconstruction also demonstrates the centrality of transnational relationships in the reconstruction of media infrastructures, as we saw in the participation of Eastern Bloc governments in the redevelopment of media and film infrastructures. Cambodian international relations were strongly tied to media infrastructures in the PRK period, as the government built infrastructure in cooperation with Eastern Bloc countries, particularly the Vietnamese and USSR governments. Vietnamese, Russian, Czech, and German goods and people traveled transnationally, and Cambodian media technicians received training from these foreign states, sometimes traveling to do so. This period foreshadows the continued non-Western cooperation in media governance and infrastructure build-out, which we still see as one of the key issues in the contemporary geopolitics of technology in Cambodia.

The political nature of the infrastructural work compels me return to the subject of the partiality of the oral histories presented here. These men were all state workers and their shared perspectives, as well as their work, were conditioned by poverty, censorship, foreign interference, and violence. Their stories also have a backdrop of inequality and elitism. These men who had access to technologies were relatively better off than other Cambodians during this period. There were also important gender dimensions and inequalities to their media reconstruction: in each story, men were working more closely with technical apparatuses than their wives, who supported them with other forms of work.

This chapter also revealed the affective nature of the work of people like Bou Vannarith, Mao Ayuth, Ka Toy, and Loak Kamala as they creatively reconstructed media infrastructure starting in 1979. They acted in courageous, passionate, and resourceful ways to repair the film and radio sociotechnical assemblages within Cambodia. Working alongside other forms of commemoration and reckoning, their infrastructural restitution became a mechanism for emotional expression for them, and perhaps for the listeners of the content they disseminated.[61] The creative production and dissemination of radio and film content processed the violence of the Khmer Rouge. Recovery also manifested in the (re)construction places of entertainment and their corresponding happy times—at home movie theaters, outdoor movie screenings, and places to listen to radio songs and stories together. Because of the active war happening around them, though, absolute recovery was delayed and media work continued to be violent.

Observing the infrastructural restitution of the PRK period from the vantage point of contemporary Cambodia involves holding together many, and sometimes seemingly conflicting, truths at once. The work was creative and often highly affective; the results were often beautiful, nostalgic, and commemorative. Yet the political messaging in the content produced during the time—such as the 1984 "Preparing for Cambodian New Year" program I described at the beginning of this chapter—can be disorienting for listeners who did not live through the period themselves and needs to be put into historical perspective. The opening ethnographic description of the participatory art event "Listening from the Archive" I began this chapter with gives a sense of the mixed feelings and curiosity that contemporary young people have about this period, and the value in revisiting media artifacts from it. These are all themes that I will pick up again in the second part of the book. The PRK period, a lesser told history of Cambodia, is also crucial to understanding the transnational relationships and governance of contemporary media culture, including internet culture. The infrastructures the PRK put in place still persist, and many people who hold positions of power today came of age during the PRK and are being (to some extent) challenged by new cultures of youth and digital tools.

INTERLUDE: PEACE TALKS

In December 1987, as the Cold War was winding down, Prime Minister Hun Sen of the PRK and Prince Norodom Sihanouk of FUNCINPEC met in France and agreed to begin peace discussions to end interparty violence. On April 29–30, 1989, the National Assembly of the PRK changed the national name "People's Republic of Kampuchea" to "State of Cambodia" (SOC). The national flag, anthem, and military symbols also changed. Buddhism, which had been only partially reestablished during the PRK, was reintroduced as the national religion.

The first session of the Paris Peace Accords occurred from July 30 to August 30, 1989, and brought together the four competing Cambodian parties—but no agreement was reached. The PRK proposed that if Prince Norodom Sihanouk returned to a governing role as king, the Vietnamese soldiers in collaboration with the PRK battling Sihanouk's FUNCINPEC army would withdraw. With the reduction of Soviet aid to Vietnam, this agreement made sense for Vietnamese forces, too, who withdrew from Cambodia between September 1989 and mid-1990. With this agreement in place, the four parties met from October 21 to 23, 1991, and signed the Paris Peace Agreement. The key goal of the agreement was to establish national reconciliation.[1] The People's Republic of Kampuchea party then changed its name to the Cambodian People's Party (CPP).

The Paris Peace Agreement gave the United Nations a mandate to create the United Nations Transitional Authority of Cambodia (UNTAC) in order to support a "free and fair" election in May 1993 and to give international financial support for the "rehabilitation and reconstruction of Cambodia." Six months after the signing of the Paris Agreement, in March 1992, UNTAC arrived in Cambodia to an environment of extreme tension and hostility. There were many challenges right away with the UNTAC mission. It brought 22,000 military and civilian personnel drawn from over 100 countries and cost the international community in excess of US$2 billion for a span of eighteen months. According to Heder and Ledgerwood, academics who worked at the Information/Education Division of UNTAC, UNTAC staff positions were financially lucrative, and "more than a few" UNTAC staff made more money than they had in the previous five or ten years.[2] In poverty-stricken Cambodia, the huge influx of foreign workers brought distorted wealth dynamics and increased the vulnerability of Cambodians living near or below the poverty line.[3]

An election in Cambodia at this volatile time was a difficult proposition. The country was used to a "winner-take-all" patronage system, where party leaders helped facilitate the livelihoods for the people who supported them.[4] The coalition between Hun Sen's CPP and Sihanouk's FUNCINPEC, and the cooperation of the other parties, was on shaky ground due to deep-rooted distrust. The Khmer Rouge (which was a minority party) and the CPP (currently in control) were particularly resistant to help the UNTAC mission since they had the most to lose in an election.

There was widespread cultural friction in Phnom Penh from the beginning of the UNTAC mission due to popular local perceptions of a condescending attitude from many UNTAC staff. Many UNTAC employees blamed Cambodians for the violence of the Khmer Rouge without recognizing the reconstruction that had already occurred in the 1980s. The mission also broadly marginalized local skills and knowledge. Though some UNTAC staff were motivated by utopian ideals, they also widely assumed "hierarchical relationships between the helper and the helped."[5] Instead of partners, Cambodians were treated as targets of the peace-building operation.[6] According to Heder and Ledgerwood, "UNTAC at times seemed pervaded with the condescending belief that the 'Cambodians' were incapable of anything unless UNTAC held their hands and walked them through it."[7] For many Cambodians, "UNTAC was seen as one more occupying army."[8]

For Cambodians hired to work for UNTAC, their role often mimicked the long-standing hierarchical structure of political/social patron-client networks. There was a perception that forming relationships with an "obviously rich and powerful friend enhanced one's life chances to obtain money, power, and influence."[9] Cambodian UNTAC employees also often kept mixed alliances and would work for UNTAC while maintaining relationships with another political party.

Many foreign UNTAC staff understood little about Cambodian culture, and the mission overall did not promote culturally sensitive behavior. Some UNTAC employees acted particularly badly, and to many Cambodians, "UNTAC was seen with horror as a horde of drinking, whoring, half-naked drivers who ran over people and couldn't care less."[10] The incidence of HIV increased dramatically in Cambodia during this time, likely fueled by relations between UNTAC workers and Cambodian sex workers.[11]

Though there were bad actors within the large number of UNTAC employees, there also existed a number of idealistic workers who cared greatly about improving life conditions in Cambodia. The information division of UNTAC specifically hired academics who had significant knowledge of Cambodian language and culture. Judy Ledgerwood, for example, had learned how to read and write Khmer at Cornell. Anne Guillou was an anthropologist and was so fluent in Khmer language that she could act as a radio host. Both went on to write some of the most important scholarship of 1990s Cambodia. Steven Pak returned to Cambodia after nine years of living as a refugee in California and developing a career in journalism. He told me in an interview that he was honored to come back to try to improve the media sector and promote democratic processes in his home country. He emphasized that while UNTAC did not do things perfectly, it tried its best under difficult conditions. The next chapter begins to address how the media infrastructures constructed and repaired in the first two chapters lived on in the transition to the contemporary political structure of Cambodia.

3

THE VIOLENCE OF DEMOCRATIC MEDIA: MEDIA INFRASTRUCTURE TRANSITIONS IN 1990S CAMBODIA

> UNTAC officials thought up a novel idea, at least they thought it was. They announced that some road works had to be done, and all those who joined the work brigade will receive radios as a token of appreciation. On the first day of the project, there were more workers than radios available so distribution was made on a first come first serve basis. On the following day, the officials found the UN office broken into, and things stolen, obviously by those who did not get radios. Later in the afternoon, one man came in with a hand grenade demanding a radio. They hastily found one for him. Then, more people turned up with mortars, grenade launchers and AK-47s. The UN closed shop.
>
> —"Grenades in Exchange for Radios," *Cambodia Times*

This chapter focuses on the ways that international institutions impacted media infrastructure reconstruction in Cambodia during the period of opening of markets and democratization beginning from the Paris Peace Agreement. Along with other vast political and economic changes, UNTAC opened a radio station that departed from state control. This opening of this radio station then ushered in commercial and independent (nonstate) media voices, including, later in the decade, early connections to the internet. Though these internet efforts made a promise of "magic eyes and magic ears," they were, in practice, inaccessible to many and built under conditions of deep inequality.

The leading question of this chapter is: How did media infrastructures change in Cambodia when the media sector transitioned to a more democratic and neoliberal environment? The first section of this chapter concerns the construction of the Radio UNTAC broadcasting station (July 1992–September 1993), the first nonstate media outlet since the Khmer Rouge came to power and the first broadcasting station ever created by an

Figure 3.1
Radio UNTAC giveaway program, from the United Nations Archives, Unique Identifier UN7746220, 10/01/1992

international governance organization. I describe the entwined history of media infrastructure and a history of violence by illustrating the cultural frictions that emerged during the process of (re)constructing the material, social, and ideological infrastructures of nonstate/independent and capitalist media infrastructures.

My first argument is that UNTAC's work on the radio catalyzed violence largely because it reacted naively to the long-standing links between politics and media in Cambodia. As described in the previous two chapters, media have long been tightly linked to a history of violence, foreign interference, and political power in Cambodia. The history of foreign interference in media showed up materially for UNTAC. Before the construction of a new radio station a month before the May 1993 election, they made do with legacy technologies from a long line of foreign governments who had contributed to the Cambodian media ecosystem since the USIS interventions. When UNTAC entered the Cambodian media sector, it became another one of those foreign interfering forces and participated inadvertently in a combustible environment, instigating further violence. For instance, UNTAC's dissemination of personal radios became riddled with violence when multiple offices were attacked in provinces. Violence was also present in radio content. The introduction of democratic media allowed for multi-party representation on radio programs, and several of the parties used racist and aggressive language in their campaigns.

The second section of the chapter describes broader post-UNTAC media transitions and shows how the economic developments that accompanied democratizing efforts affected Cambodians unevenly, drawn out through interviews and archives about the early Cambodian internet. I focus specifically on the 1999 construction of the Cambodian portion of Alcatel fiber-optic cable, the first fiber-optic cable laid in Southeast Asia. I work in part from Rithy Panh's film *The Land of Wandering Souls*, which documents the migrant worker families who installed the cable and encountered remnants of the war period along the way. This film illustrates the trauma and extreme poverty experienced by the migrant workers who constructed new globalizing telecom infrastructures. These workers needed to dig up not only bombs but human bones out of the ground to lay the fiber-optic cable.

My sources for this chapter come from the University of Wisconsin's collection from Radio UNTAC, delivered to Wisconsin by the UN and uniquely held there.[1] I also use information from *Radio UNTAC*, a memoir by Zhou Mei, a Singaporean journalist. This memoir gives insight into the material constraints, cultural challenges, and achievements of Radio UNTAC but comes from a highly racialized perspective, as I discuss. I also use archives of the Foreign Broadcasting Information Service (FBIS online archive), the *Cambodia Times* archives from 1992 to 2000 at the National Archives of Cambodia, and miscellaneous documents about the telecom industry and the internet held at the National Archives of Cambodia. I interviewed Judy Ledgerwood, Steve Heder, Jeffery Heyman, and Steven Pak, who all worked at UNTAC in the Information Division or Radio. I also interviewed Norbert Klein, who worked for Open Forum of Cambodia beginning in 1990 and started a peer-to-peer internet platform in Phnom Penh, and Chakrya Moa, the telecommunication regulator of Cambodia since 1997.

1 RADIO UNTAC

The Cambodian media landscape dramatically changed during the UNTAC transition. In April 1992, the State of Cambodia (SOC) parliament passed a media law that gave the SOC control of media. Just afterward, the UNTAC Information/Education Division prepared an opposing "media charter" that provided a legal framework for a free press operating in the administrative zones of all four factions. Although under shaky legal grounds, it took precedence over the SOC law.[2] This "watered-down" set of media guidelines determined what was and was not appropriate for the four parties to broadcast and publish.[3] It determined that political parties would get equal time on UNTAC media sources and could not broadcast political messages during the elections. This charter also gave unprecedented room for independent media to exist.[4]

Apart from Radio UNTAC, as discussed in the previous chapter, the major media outlets in Cambodia in 1992–1993 were the Voice of the Cambodian People (the State of Cambodia Radio); the Voice of the Great National Union Front of Cambodia; the renamed Khmer Rouge or PDK radio, operating clandestinely from Pailin on the Thai border; and a new FUNCINPEC radio from the Thai refugee camps.[5] The "Voice of the Khmer People," the joint FUNCINPEC-KPNLF station discussed in the previous chapter, had started broadcasting into Cambodia from a stronger transmitter in Chiang Mai.[6]

Much of the rhetoric on the radio was deeply partisan during the UNTAC period, as was the case in the PRK period. Anti-CPP media were harshly expounding racist messages against the Vietnamese while illustrating the CPP-Vietnamese alliances.[7] As Ledgerwood and Heder explain, the propaganda offered by all the political parties dehumanized other political parties and foreign people, including UNTAC staff and—most prominently and viciously—Vietnamese ethnic groups in Cambodia (both ethnic Vietnamese Cambodian citizens and Vietnamese citizens). They were often dehumanized through labeling them "non-Khmer," "traitors," "genocidists," or "criminals."[8] Labeling people as non-Khmer was so powerful in this context because, according to Ledgerwood and Heder, there was an "idea that Khmer culture might disappear . . . or an overwhelming feeling, indeed obsession, that Khmer culture is being lost." Ledgerwood and Heder point to the continual war, the loss of life during the Khmer Rouge, the emigration of hundreds of thousands of Cambodians, and the Vietnamese occupation as the driving forces for this cultural fear. As in the PRK period, this fear was coupled with a fierce competitiveness about which party was most legitimate to rule Cambodia.

In this highly charged political environment and in an unprecedented move for an international governance mission, UNTAC invested in a broadcasting facility under the Production Unit of the Information/Education Division. It established a major radio station—Radio UNTAC—as well as a smaller TV station, some newsletters, and graphics work. Tim Carney led the Information Division of UNTAC and was responsible for developing the radio station. With the help of Steve Heder, he slowly staffed the project with a mix of international journalists, Cambodian "re-pats," and Cambodian staff, some of whom had prior experience at the National Radio.[9] Steve Heder and Jeffrey Heyman, in interviews, both describe the difficulty and ad hoc nature of hiring for the station. The Information Division hired a mix of academics who were familiar with Khmer language and culture and people experienced in broadcast media. Jeffrey Heyman was a broadcaster from the United States who was just traveling through the region on vacation when he was hired. In an interview, Heyman described the early days of the radio station as a "wild scene."

Radio UNTAC started broadcasting in July 1992 with two half-hour Khmer programs per week broadcast on the state station.[10] The radio staff taped the programs and drove them to the State of Cambodia transmitter in Stung Meanchey (a thirty-minute drive) for broadcast.[11] UNTAC shared the transmitter with the State of Cambodia until the SOC started using a new Russian-funded transmitter in early 1993. The UNTAC/SOC transmitter was the same Philips 120-kW radio transmitter and antenna mast left over from the Chinese development assistance in 1959–1963 and used throughout the PRK period.[12] It was difficult for UNTAC to find spare parts for the old transmitter, but the radio team ultimately found and shipped parts from Australia.[13] UNTAC officials paid two Russian technicians to "patch up" the antiquated equipment regularly. Sometimes the power had to be drastically reduced from 120 to 70 kW. The transmitter (which Mei calls an "innocuous junk-piece") occasionally broke down despite all their upkeep. UNTAC also borrowed from the SOC a 450-kW generator, which was already at the Stung Mencheay site, to run the transmitter.

In November 1992, Radio UNTAC started broadcasting on its own frequency, MW 918 kHz. UNTAC also aired Radio UNTAC from Bangkok on the Voice of America station to reach the borderlands where the MW 918 kHz station could not reach. In early 1993, UNTAC built three medium-wave relay transmitters in Sihanoukville, Siem Reap, and Stung Treng, which were running by April 1993 (only a month and a half before the election). In order to build these relay towers, UNTAC had to do extensive demining around the sites of the transmitters (which Jeffrey Heyman told me was very "logistically complicated"). With these relay towers, the radio was able to reach 90–97 percent of the geographical region of the country.[14]

On April 5, 1993, Radio UNTAC moved into a newly renovated UNTAC Complex, which was funded by the United States and cost over $3.1 million (around $1.9 million for equipment, $1.1 million for installation, and $40,000 for three months of maintenance). Their new studio was air-conditioned and included six custom-built studios (not all were used). Each of the six studios came with stereophonic audio-mixing consoles, a digital harmonizer, a digital reverberator, a digital logger recorder (which never worked), loudspeakers, and broadcast microphones (which were never used). Radio producers used a portable digital audio tape (DAT) recorder that needed special batteries and DAT tapes (which luckily were already being used regularly before the upgrade). There was a telephone system in the studio, which was broken by lightning almost immediately after installation.[15] After moving into the new studio, broadcasts could be played live and sent by microwave to the transmitter in Stung Mencheay, and from there to the relay towers. The live broadcasts started with three three-hour segments aired during peak listening periods for a total broadcast day of nine hours.[16]

Content on Radio UNTAC included news about the election, pop culture, and public "betterment" programming. Radio UNTAC had a signature opening and closing theme song, borrowing a wedding song in Cambodia, traditionally played during an important knot-tying ceremony.[17] In April and May 1993, an editorial series was devoted to introducing each party running in the election to the people. The radio also broadcast political roundtable talk shows and debates of the constituent Assembly—"a first for a radio in the country."[18] Each party had two five-minute slots per week of UNTAC airtime. Other content included news and statistics about the election, interviews with key figures in the peace process (both UNTAC officials and local political leaders), dialogues about the elections in skit form, and dialogues on human rights including the right to information, to opinion, and to vote. The skits often emphasized the secrecy of individual votes and the principles of democratic government. The content also included public health dialogues about, for example, preventing malaria and dengue fever.

Figure 3.2
Radio UNTAC flier, University of Wisconsin. Source: UNTAC Collection

Cambodian traditional and pop music songs from the 1960s were interspersed with the talk radio.

Radio UNTAC featured a hugely popular song request and letters from listeners program.[19] The station received up to 1,500 listeners' letters a day. The *Cambodia Times* comments on the novelty and excitement created by the program: "Sixty per cent of [the letters] were asking for their favorite songs, the rest, some anonymous, had all kinds of questions and gripes." The uncensored content challenged the SOC control. The *Cambodia Times* continues, "The frank replies [on air] often annoyed authorities, who accused Radio UNTAC at times of violating 'local tradition.'"[20]

At first, before programming went live, programs were written in English and French and then translated into Khmer, then read and taped and driven to Stung Menchey, representing a huge amount of work per program. Foreign producers who spoke Khmer vetted the final Khmer broadcast "to make sure that there was nothing misinterpreted or biased in the Khmer translations." With time, programs were planned and presented only in Khmer and broadcast live.[21]

In order to provide access to the radio programs, UNTAC facilitated a radio giveaway program starting in September 1992. Though accurate statistics on radio ownership are not available, Radio UNTAC staff recognized that because of widespread poverty, many families and communities did not have access to a radio and batteries.[22] Between September 1992 and April 1993, UNTAC received a donation of 347,804 radios, 849,400 batteries, and 1,000 radio cassette recorders from Japan. The bulk of these were second-hand from the Soka Gakkai company and the Social Democratic Party of Japan; however, the Japanese government provided, as part of the total, 40,000 new radios, 396,000 batteries, and 1,000 radio cassette recorders.[23] "The first radios were accompanied with batteries. When the batteries ran out, UNTAC staff showed recipients how to connect them to motorcycle and car batteries for power."[24]

The logistics of disseminating these radios were a major challenge. Alex Huber, who worked for the UNTAC Information/Education Division, was responsible for disseminating the donations but he did not have access to easy or direct transport, warehousing, or human resources. Zhou explains that he accomplished the task through what he calls "his persuasiveness, dogged determination and perhaps when deemed appropriate, coercion."[25] Soldiers of the Ghanaian Battalion helped. Huber developed processes for warehousing, loading and unloading, processing, and testing the equipment before transporting the radios to twenty of Cambodia's twenty-one provinces.[26] He flew the first batch (of approximately 50,000 radios and over 100,000 batteries) to Phnom Penh via Bangkok and then distributed them by road. He later shipped containers of radios and batteries via truck to Sisophan, a town in Benteay Menchey in the northwest of Cambodia. Sisophan received forty-four 40-foot containers of donated radios. He then shipped these by car to the western provinces. The last batch of radios came from the government of Japan and were new radios, compactly wrapped. Alex arranged for these to be flown by helicopter to recipient provinces.[27]

Giving out these radios when they arrived in recipient communities was also not an easy task. The radio distribution became violent. Zhou says, "UNTAC had inadvertently provoked some ill-will. There were those who became angry, incensed, when they found that they were not among the recipients. The desire for possession was so overpowering that at times, people became ugly. There were incidents of near riots in UNTAC compounds where the radios were stored. But, all that should not distract one from the

Figure 3.3
Radio UNTAC giveaway program, from the UNA, Unique Identifier UN7746222, 08/01/1992

accomplishment. Without the tools to receive UNTAC's messages, Radio UNTAC's work would have been for naught."[28] The *Cambodia Times* reports the violence of this distribution more dramatically, explaining that it led to people showing up with mortars and hand grenades. An article from June 1993 explains:

> UNTAC officials thought up a novel idea, at least they thought it was. They announced that some road works had to be done, and all those who joined the work brigade will receive radios as a token of appreciation. On the first day of the project, there were more workers than radios available so distribution was made on a first come first serve basis. On the following day, the officials found the UN office broken into, and things stolen, obviously by those who did not get radios. Later in the afternoon, one man came in with a hand grenade demanding a radio. They hastily found one for him. Then, more people turned up with mortars, grenade launchers and AK-47s. The UN closed shop.[29]

This story is a clear example of the links between violence and the reconstruction of media infrastructure. UNTAC underestimated the strong historical links between political power and media in Cambodia, which the first

Figure 3.4
Radio UNTAC giveaway program, from the UNA, Unique Identifier UN7746220, 10/01/1992

two chapters of this book illustrate. This political power combined with an active war, poverty, and inequality led to a volatile environment that UNTAC's radio dissemination clearly inflamed.

By May 12, 1993 (ten days before the election), Radio UNTAC was live for fifteen hours per day (from 5:30 am to 8:30 at night).[30] During this highly active time, UNTAC staff traveled extensively to the rural provinces, broadcasting live and distributing audiocassette tapes that explained how to vote. They also distributed videos "which UN personnel surreptitiously slipped in between 'two Rambo movies' on sets in local cafes."[31] During the election between May 23 and 28, 1993, many of the Radio UNTAC producers went to rural provinces to report on the election process and broadcast ballot statistics twice a day live. Steven Pak says that he would send a transmission of live activity from the provinces via a Motorola cell phone to the Phnom Penh studio. The radio station also boldly reported the consolidated vote counts live from Phnom Penh, angering the SOC authorities.

Radio UNTAC was very popular and staff have widely called it a success. Voter turnout was nearly 90 percent, and Radio UNTAC staff and the

Figure 3.5
Radio UNTAC reel

Cambodia Times gave the radio station credit for this high turnout.[32] In 1993, with a Cambodian population of approximately 10 million, UNTAC estimated six to eight million daily listeners and up to 11 million in total audience for Radio UNTAC.[33] John Marston, who was conducting research in Cambodia during this time, explains the significance of Radio UNTAC. He says that it was the first radio station to act like a corporate radio station. He writes,

> The international technical expertise of the broadcasts was no doubt an element of what generated a mood of excitement about the radio and made it more attractive than its competitors . . . it is fair to say, though, that beyond the success it achieved because of its capital and technical expertise, Radio UNTAC was also successful because it consciously cultivated a sense of connection with the audience by, among other things, reading letters from listeners on the air and playing request songs. Cambodian announcers were given freedom and responsibility and genuinely tried to report issues fairly. Issues of human rights abuse were dealt with frankly—probably one reason why SOC complained that Radio UNTAC was biased against it.[34]

After the conclusion of the vote, FUNCINPEC won 45 percent of the votes and the CPP won 38 percent. The CPP, however, negotiated against the results and shared power with FUNCINPEC. In June 1993, Radio UNTAC received death threats. A handwritten letter was delivered to the radio complex that said that two CPP leaders wanted them "exterminated."[35] Radio UNTAC also received death threats over the telephone: "Stop broadcasting or die." The *Cambodian Times* reported, "Government tanks (from the former regime) appeared at the corner of the street. We [UNTAC staff] called for 50 Ghanaian Blue Berets as reinforcements, and the programme went right on, says Jeffrey Heyman, the technical director."[36] In an interview with me, Heyman confirmed the fear and defiance that the staff all felt at the conclusion of the election.

The CPP officials disagreed with the "success" rhetoric of Radio UNTAC. Hun Sen said in June 1993, regarding Radio UNTAC's broadcasts of the election results: "If Radio UNTAC wanted to be like gasoline poured on the fire, please be responsible for your security yourselves . . . there was a great demand to demonstrate against Radio UNTAC at its station, but it was prevented. . . . Now, if Radio UNTAC wants to become gasoline pouring on the fire, please go ahead."[37] Other CPP officials claimed that news coming from Radio UNTAC biased the election so much to cause irregularities that might invalidate the election.

On June 3, 1993, Kheiu Kanharit, spokesman of the government of the State of Cambodia, said on the Kampuchea Radio Network that they wanted the "people to remain calm" and asked Radio UNTAC to hold back from broadcasting the election results. "The SOC position is that we will respect the results of a genuinely free and fair poll because it represents the will of the people, but now we should tackle some irregularities." He continued that UNTAC was unfair in how it enforced the media rules. UNTAC "forbade the SOC radio and television from broadcasting political issues," but the "VOA continued its offensive broadcast against the SOC. The FUNCINPEC television also continued to introduce its leaders without being banned or warned by UNTAC." He had a number of other grievances and implied that Radio UNTAC compromised the sovereignty of Cambodia and the fairness of the election.[38]

Radio UNTAC wrapped six months after the conclusion of the election and UNTAC gave the radio equipment from its main offices in Phnom Penh and its relay stations (in Stung Treng, Siem Reap, and Sihanoukville) to the government of Cambodia, which was still using 1960s equipment except for the new transmitter built with Soviet help. Bou Vannarith (mentioned in the previous chapter) showed me some of the UNTAC-donated

Figure 3.6
Vannarith at the National Radio studio, using UNTAC equipment, September 2017

equipment, which was still being used by the Cambodian National Radio station as of 2017.[39]

In the years after the departure of UNTAC, Cambodian political mechanisms returned largely to their status quo from before the election. Though FUNCINPEC won at the polls, they were relatively weak militarily so could not maintain power. By 1997, Hun Sen ordered a coup including a grenade attack in Phnom Penh. Shortly after, Hun Sen was elected sole prime minister in 1998 with Prince Ranariddh as president of the National Assembly.[40] According to Ledgerwood and Heder, UNTAC as a whole failed to introduce democracy because "the social, cultural, and institutional bases for the emergence of a democracy movement based on an urban middle class or urban civil society were virtually nonexistent."[41] The "democratic transition" was only a one-off exercise: the May 1993 election.

Apart from the "success" or "failure" of Radio UNTAC, my analysis of the program is that it was an ahistorical foreign intervention into a sector that was deeply political and had a long history of conflict. Some of this history left material traces. Foreign governments, including the US, Vietnam, and China, had previously invested in material infrastructure that supported their interests in Cambodia. UNTAC's infrastructure worked from the material legacies of these prior interventions. The material remnants of war such as mines around media infrastructure technologies were indicative of the short jumps between violence and media in this context. In addition, since the Sangkum Reastr Niyum period, there have been strong links between a perception of national legitimacy and control over the radio. Given this history, the violent language in radio content was not altogether surprising. UNTAC's lack of historical sensibility was exacerbated by the broader perception of UNTAC—accurate or not—that many UNTAC workers were new to Cambodia and were culturally insensitive. If UNTAC had regarded this history seriously, they might have been more cautious in building out this program and avoided the kinds of violent flare-ups that the program saw materialize.

This ahistorical character marks the media construction work of UNTAC as distinctly not a kind of infrastructural restitution. Some of the early patchwork moves they made to repair radio equipment take on some of the air of resourcefulness that we saw in the infrastructural restitution of the previous chapter. The people participating in the construction of Radio UNTAC, however, did not emphasize the past (the good or the bad) but instead focused only on the future, on disrupting the current political climate and making a new democratic system. They had a noted absence of historical perspective in the work they were doing, perhaps in part because many of them were not from Cambodia and had no former connections to Cambodia. This case gives a clear demonstration of the value of historically informed international development.

The contrast between this work and infrastructural restitution, however, also suggests a reflection on the conservative impulses of infrastructural restitution. Infrastructural restitution often looks backward romantically and in so doing can harden historical inequalities. For instance, a sense of cultural nostalgia is tied to a long-standing anxiety in Cambodia about "losing" Khmer-ness due to violence or foreign interference. Romanticizing the past can be tied to problematic quests for a pure past Khmer-ness, which are often tied to a desire to maintain long-standing power structures. The violent, racist language so common to the parties' radio content in the UNTAC and PRK periods is—in part—the product of a fight for legitimacy, with different parties' claims often based on arguments that they are most connected to a past "true" Khmer ruler. These calls to history can include the exclusionary, xenophobic, or racist language that was so common in the election propaganda.

2 BROADER POST-UNTAC MEDIA CHANGES

Putting aside whether or not the program was well conceived, the process of building a democratic media infrastructure during the UNTAC period ultimately transformed the media landscape in Cambodia. In the midst of Radio UNTAC opening and wrapping up, many other nonstate, nonparty media became available from Phnom Penh. These independent media provided more and different kinds of information to Cambodians, and gave a sense that a greater kind of change was taking place.[42] Marston wrote in 1994, "Phnom Penh media are no longer socialist but now function in relation to a free-market economy, although they do not operate like those in most Western countries."[43] This opening of media meant that the 1980s geographical division between the Phnom Penh media and a media of resistance ceased to exist during the UNTAC period.

Economic changes also contributed to the growth of the media sector. The national economy grew 8 percent in 1992 and foreign investors started pouring in after the election. Partially due to this economic growth, television reached an inflection point in the fall of 1993. The Thai-owned IBC was the first commercial television station to open in late 1992. Access was still relatively limited (there were only 300,000 TV sets in the country) and it didn't represent a major information source for most Cambodians.[44] IBC projected a (hoped-for) market of 6 million Cambodian viewers, but its transmitter reached only 100 kilometers from Phnom Penh; they delayed development of a Kampong Cham relay station and ultimately used only satellite transmission.[45] IBC, however, started making new Cambodia-specific programs, including one called *Cambodia's Reunion Programme*, which showed the reunions of Cambodians who had been separated during the Khmer Rouge and spread out to different parts of the world. IBC transitioned from thirteen hours to twenty-four hours of programming in 1994.

The 1990s saw a huge increase in independent filmmaking. Phnom Penh hosted a film festival in 1990, signaling the beginning of a return to a private film industry. Ka Toy told me that there was a big change in filmmaking when the UN came to Cambodia and the markets opened up.

Yvon Hem (an active filmmaker from the pre–Khmer Rouge period) and Mao Ayuth made several incredibly popular films. Despite this growth in the industry, most of my participants lament that these "weren't the same" as the films from the 1960s.

Cinemas also changed dramatically. They stopped using film projectors and instead started to use LCD projectors or VHS. Cinemas became more comfortable and Mao Ayuth commented that after UNTAC, Phnom Penh cinemas were air-conditioned again. Due to market forces, however, many of the heritage cinemas also started to close down in the UNTAC transition. Some were sold, privatized, or leased to the businessmen to be run as something new.[46] These changes led some PRK-era media workers such as Ka Toy to transition their work.[47]

There were also major changes to the telecom sector. The limited prewar telephone network had been destroyed during the Khmer Rouge, and new infrastructure was never built during the PRK period. By 1992 (when mobile phone technology was introduced), there were only 4,000 fixed telephone lines for a population of 9.8 million.[48] By the year 2000, four out of five telephone subscribers used a wireless phone, which was the highest ratio in the world at that time; this represented, however, only one phone line per 2,000 people.[49] The access to telephones was extremely split by the urban-rural divide. In 1996, 5 percent of Phnom Penh homes had a telephone, whereas less than 1 percent of overall households had a telephone. By 2000, 85 percent of the country's fixed lines were in Phnom Penh (though it accounted for only 10 percent of the country by population). The low access to telephones can be partly explained by cost; in the early 2000s, international calling rates remained very high.[50] According to the Council for the Development of Cambodia, private capital totaling US$131 million was invested in the telecommunications sector during the period 1994–1999. The Ministry of Posts and Telecommunications was involved as either a provider or joint venture partner in almost every telecom project, but it invested hardly any of its own money.[51]

Internet exchanges began in Cambodia around the same time as UNTAC. Many foreign workers came to Phnom Penh in the UNTAC period and directly after. According to a 2002 International Telecommunication Union (ITU) internet report, this international community first used and promoted the development of internet services.[52] One of these early email exchanges in Cambodia was the one that Norbert Klein organized in 1994, and he explained the opportunities and challenges of the early Cambodian internet to me in an interview in January 2020.

Klein first came to Cambodia from West Germany in 1990 as part of a Swiss international development organization to work for the Department of Agriculture in Phnom Penh. He explained that he established the first internet email exchange in late 1994 in order to help a colleague in the Department of Agriculture attend a scholarship program abroad. Klein established the .kh domain for Cambodian web pages. The initial system was expensive and limited in capacity, mostly used by foreigners communicating with family abroad. The 2002 ITU report corroborated Klein's report of the challenges for building internet infrastructure. It claims that these challenges included a "lack of vibrant academic community that could help nurture and sustain networking; the complexity of computerizing the written Khmer language, which hinders local application development; an extreme shortage of dial-up telephone lines needed to access the Internet; and government policies that have restricted Internet supply."[53] Klein stopped facilitating the exchange in 1996/1997, when the Ministry of Post in Australia established a satellite link that was much faster and cheaper than the telephone link. Several other internet service providers (ISPs) opened in 1997 and 1998. Klein at that time transferred the .kh domain to the Ministry of Telecommunications. He later worked for the Cambodian branch of Open Forum (later renamed the Open Institute), an NGO that was established during the UNTAC era and advocated for open communication on the internet.

In 2001, there were two ISPs: the government-owned Camnet and the foreign corporate ISP BigPond. Together they had fewer than 4,000 subscribers, and these were only in Phnom Penh and Siem Reap (an entry-level plan costs $3.99 per hour, eleven times what Singaporeans were paying). In mid-2001, two more private ISPs and telecommunications providers started operations (Camintel and MobiTel).[54] According to the ITU in 2002, Cambodia had the lowest internet penetration and the highest prices for internet in the Southeast Asian region.[55]

In the early 2000s, there were between 50 and 100 internet cafes in Phnom Penh and Siem Reap. Phnom Penh residents could take PC training courses for 1,000 riel ($0.25 USD) per hour through NGO programs. Some NGO projects addressing internet access started around this time, including access to low-cost public internet provided at the Public Interest Center in Phnom Penh established by Pan Asia Networking.

To support the growth of the internet and build a Southeast Asian market, in 1999, Alcatel, in collaboration with the Ministry of Posts and Telecommunications, built a fiber-optic cable that runs from the Thai border through Phnom Penh to Laos. The construction of this cable is the topic of *The Land of Wandering Souls* (2000), a documentary by Rithy Panh. He portrays the life of Cambodian migrant workers who travel across the country to work on the placement of the cable, connecting Thailand to Laos. The cable crosses Cambodia to connect up with the cable laid from Europe, following the path of the historical Silk Road. The documentary portrays the poverty, trauma, and hardships the workers experienced—in contrast to the promises of access and connection that early internet evangelists espoused. The film focuses on a family from Kampong Cham, who is no longer able to make enough money to eat through farming rice in their hometown and thus joined the mobile work crew. The wife/mother character, Mon, is pregnant, and her husband, Sela, has an amputated leg from an injury during the wars. Mon's family can barely manage on the wages from placing the cable.

Mon says to an older friend, "We dig every day. Sometimes his stump gets infected . . . even if it smells bad, he has to keep on digging. When he takes his limb off, even if it smells awful, I have to bear it. I'm his wife. That's our fate. He respects his wife and children, I love him for that. He has a good heart. He goes without food to feed his family." The older woman with whom Mon chats commiserates by telling Mon about her hardships during the time of Pol Pot. "I still have scars. I had to dig irrigation canals. The memory is unbearable." She then says: "Suffering is the same for poor people everywhere . . . The poor suffer too much."

As the workers dig for the cable, they start to hit things in the earth. Returning from a day of digging, Sela tells Mon that he found a bomb. "I nearly died of fright. I thought, that's it! One more strike of the hoe and it would have exploded." Later in the film, the workers start coming across bones. Mon says, "My husband found a human bone while digging . . . the only thing to do is to sprinkle yourself with holy water. . . . During the Lon Nol and Pol Pot time there were many violent deaths. When we dig here we find bones. Nobody celebrated their deaths when they passed away, so they will haunt you. This war claimed a great many deaths in this area. National Road 1 was littered with bodies. The tanks ran over thousands of men. That's how it happened. It wasn't normal." She accepts the death fields with a calm demeanor.

Three of the male migrant workers reflect on the way that the cable would hypothetically connect them to different parts of the world. One says the cable is like a "magic eye." Another says, "Magic eyes means you're here, but you can see as far as America. You can see the whole world. That's the magic eye." His friend says, "With this cable, they transmit TV. They transmit CNN, cable news, all over the world. People watch news on television." One asks, "And the magic ears?" "You're here, but you hear some in America or Canada." Another responds, "We used to talk of magic eyes and magic ears, but now we have them. Before when we wanted to send information we had to use the mail. The letters took months, years to arrive. You send your letter by the internet. You can even send your photo. It's called the internet. I don't have electricity, only an oil lamp. I often don't have kerosene for my lamp. I have to go to the Chinaman to get 100 or 200 riels for kerosene." These workers, however, would not readily get access to the types of new connectivity that they were helping to construct.

The conditions of work were poor for these workers; the payment schedules were inconsistent; and money was often needed for immediate concerns, like feeding their kids or for paying for blessings for Chinese New Year. The film depicts Mon getting an infection in her hand from gripping the digging hoe, which swelled and filled with pus. She asks the doctor to wait for a week for payment, after they finished another section of the cable. The family has to borrow money to eat.

Workers often did this work because of a dearth of other options and desperate life situations. The film shows workers discussing prior work in Thailand with miserable conditions; some talk about using drugs to work without breaks. Workers warn of the trap of Thailand and the risks of trafficking. The workers also compare this work to their wartime experiences. One worker said that he was a soldier in the war for twenty years before he became a migrant worker. When he came back home after the war, "there was nothing for me." "Because of this war . . . they already divided up the land, leaving me empty handed. So I've been wandering across the country, this land, this land of others, to feed my family. I still have a bullet in my foot and that's why I complain about pain."

Though portraying injustice and inequality, the filmmaker, Rithy Panh, shies away from unidimensional portrayals of life in poverty or stripping people of their agency. He holds a long shot of kids swimming in still water along the digging route. Families eat noodles and rice together. One man says, "If you ask who, between rich and poor people, takes the most pleasure in eating, I would answer: the poor enjoy eating. The poor? With what money? you ask. Food is always delicious. The rich forget what's good, they have too much. If a poor man has a chicken bone, he eats it and sucks the bone. A rich man doesn't eat. He tastes it to kill his boredom. He's forgotten what's good." A worker listens to Khmer oldies from the 1960s while he works. One evening the workers have a party and dance.

Despite the toil of these construction workers, the fiber-optic cable did not, at first, even connect Cambodia to the internet—it instead only traveled through Cambodia.[56] The laying of the cable foreshadows the inequality intrinsic to the political economy of the internet in Cambodia, which continued through the course of my fieldwork. Between 2002 and 2017, the internet became mainstream in Cambodia. Connections grew astonishingly fast, particularly when internet through smartphones became available. The internet, however, is, in practice, dominated by very few companies.

In this chapter, I showed that UNTAC dramatically restructured government control over media, liberalizing it and establishing new forms of media-centric violence. Many UNTAC workers, including those in the radio, were hopeful about the positive impacts of the work they were doing. UNTAC Radio successfully developed multilingual programming and gave many Cambodians access to new kinds of radio content. The radio encouraged more people to vote in the 1993 election. Yet Radio UNTAC's restructuring of media did not have the democratizing effects that many of its workers hoped for. The election itself did not fundamentally shift governance structures in Cambodia. Though the radio opened up Cambodia to independent media voices, it also ushered in new kinds of hate speech tied to historical rifts between parties and ethnic groups. The program worked naively given the historically high political tensions intrinsic to radio in this context. Working in such a fraught environment and on top of a foundation of violence, it inadvertently instigated new violence, such as the flare-ups after an unfair radio giveaway.

This violence was encoded in the materiality of its infrastructure. There were three particularly value-laden pieces of infrastructural material in the Radio UNTAC project. First, the transmitter in Stung Menchey, developed first by Chinese funding in 1959–1962, was a material legacy of long-standing geopolitics of the Cold War that UNTAC made do with. Second, in order to build the relay stations, UNTAC had to demine the ground, working on post-conflict land. Third, the donated radios (and the logistics of their giveaway) highlighted inequalities, cyclical violence, and deep intergroup resentments. In each of these material things, violence was symbolically and/or literally embedded.

I juxtaposed the violence of UNTAC's establishment of a democratic media infrastructure with the growth of telecom and internet business developments of the 1990s and 2000s. The period ushered in new exploitative labor conditions for media workers following rapid globalization and market liberalization. Like with Radio UNTAC, the foundation of these new developments also included Cambodia's history of violence. *The Land of Wandering Souls* gives striking images of the unexpected material remnants of a history of violence. Rithy Panh's portrait of migrant laborers putting the fiber-optic cables into the ground demonstrates how bones and bombs remain trapped in the ground; violence is part of the geography of media. This film demonstrates how reconstructing new media infrastructure unearths historical trauma, and does so unequally. As the cable snaked across the country, the workers putting it into the wounded earth struggled to get by and were themselves left out of the national turn to global internet and communication technologies.

These post-UNTAC media interventions foreshadow the contemporary media landscape in Cambodia and provide some warnings. The Radio UNTAC project has resonances with contemporary information and communication technologies for development (ICTD) projects aimed at democratizing, liberalizing, or making accessible new technology programs or products. It provides a concrete example of how interventionist media projects, even under the best of intentions, can create violence, particularly when not attuned to the nuances of historical context. *The Land of Wandering Souls* depicts workers laboring on a fiber optic cable who never receive its access to the internet. This documentary predicts the ways that global media companies continue to serve different global markets unequally. Workers in the Global South often provide the backbone for global internet infrastructures through labor, resource extraction, or waste disposal, but their access to devices, networks, and software is often still limited in their markets. Companies also provide lackluster services to users in the Global South; for instance, they often provide minority language groups with insufficient support for content moderation and translation.[57] These differences in media quality and availability exacerbate global inequality and put into motion further violence.[58]

In part II of this book, I transition to how the reconstruction work that began in the early post–Khmer Rouge period continues in contemporary Cambodia. I offer four contemporary examples of infrastructural restitution, the creative reconstruction of media infrastructures, demonstrating new dimensions of its goals for reconciliation and political action.

II CONTEMPORARY MEDIA RECONSTRUCTION

4
MEDIA RUINS

The architectural historian Loak (Mister) Sok and I walked down an old colonial street in Battambang to visit Moeun Chhay, a painter and musician, in September 2017. As we walked up to his home and studio, he was painting a small figurine of a cyclo on the sidewalk outside. Some kids sat around him, watching him paint. Moeun Chhay was famous for being one of the few people still alive who painted posters for 1960s films. He grew up in this neighborhood near many of the old Battambang cinemas and hung out in his teenage years with older people working in the film industry, who taught him how to paint. He says, "Back then, when a film stopped running, I would take the canvas down to the river and wash the powder colors off and repaint a new poster for the new film." None of his original posters now remains intact.

In the back of his home was the old Maek Cha cinema. Moeun Chhay survived the Khmer Rouge period by telling the commune leader that he played chapei, a classical Cambodian stringed instrument. He played music for the Khmer Rouge throughout the regime. After the Khmer Rouge, he worked for the Ministry of Fine Arts, which secured this house next to the old cinema for him. Moeun Chhay kept the key to the cinema and opened it to me and Loak Sok. As we walked in, Moeun Chhay and Loak Sok explained to me that in the 1960s, Maek Cha was built as a theatrical hall. From 1970 until 1975, it was used as a cinema. During the Khmer Rouge regime, it sat unused. In the 1980s, it was used as a theatrical hall again to show plays about the Khmer Rouge.[1] Loak Sok and Moeun Chhay told me that there are no remaining records of these plays—they were never written down. Now the theater again goes unused.

Maek Cha is a cinema as *media ruins*: the seats are broken, the bricks holding up the screen are crumbling, and the paint highlighting the 1960s

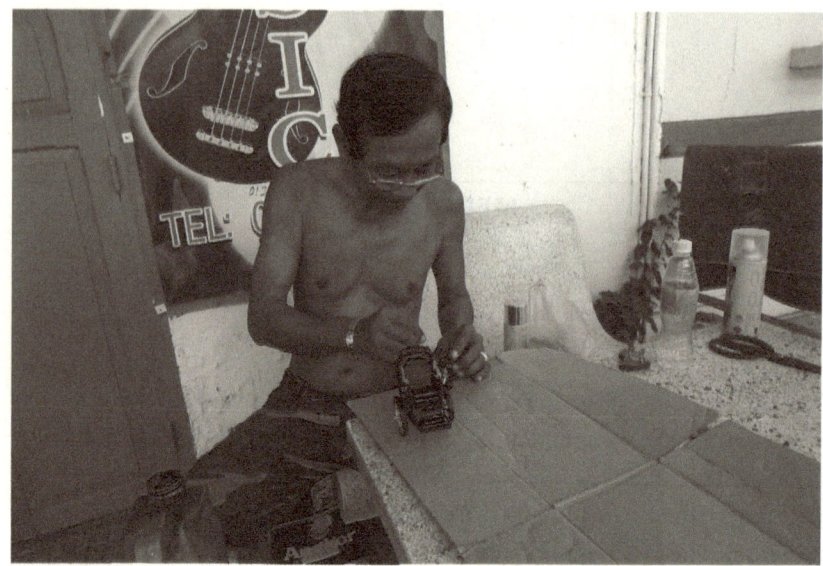

Figure 4.1
Moeun Chhay outside his studio, Street 2½ Battambang, Cambodia

Figure 4.2
The remains of Maek Cha cinema, Battambang, August 2017

modernist craftsmanship is chipping. The wooden roof has holes that let light in. The light streams through the air, illuminating particles of dust. The gray concrete walls are stained with dirt, smoke, and tar. A ripped white cloth stands in for a screen at the front of the hall, gently rippling with drafts of wind. *Media ruins* are modernist and technological spaces in decay, inhabited by memories and ghosts, including the ghostly presence of historical media itself. They have a complex and contradictory affective resonance. Though media ruins bring to mind death and cultural loss, they also elicit powerful happy memories of entertainment, like cinema-going.

To illustrate what I mean by media ruins, I describe in this chapter the work of Roung Kon (the Khmer phrase for "cinema"), the group of architects who find, survey, and exhibit heritage cinemas of Cambodia, built before the war. Through this case, I highlight how infrastructural restitution can act as a kind of hidden script of political action. I explain the context of Phnom Penh in 2017–2018, including urban change and development as well as increased political repression. Given this context, Roung Kon's patient and careful documentation of heritage cinemas and the dissemination of their knowledge—their work of infrastructural restitution—was a form of social and political action. Roung Kon praised and mourned the flourishing arts and space of the cinema from before the crisis of the Khmer Rouge in order to encourage public support for the arts and sustainable urban development in contemporary Phnom Penh. Their work also suggested an alternative to information control. I describe, too, the independent arts space where they worked, which acted as an invisible infrastructure of independent thought, arts, and interpersonal trust.

The Roung Kon team used a suite of digital tools, including AutoCAD and Google Maps, to find, research, model, and disseminate information about cinemas such as this one. For the members of Roung Kon, celebrating these sites was critical for their generation; they were important social and cultural spaces during the extraordinary time of the arts in Cambodia after independence from France (1953) but before the worst of the civil war (1970–1975) and Khmer Rouge genocide (1975–1979). Kagna, one of the members of the group, explained that "old cinemas are symbols of meaningful memories of entertainment and the arts." The mystery of the memories around the cinemas drew Roung Kon in. "Most of the cinemas are gone but we can see some structures that still exist. We want to know more about them," another member of the group, Daro, said. The group members also

Figure 4.3
The remains of Cinéstar Cinema in Phnom Penh, August 2017

told me that this project became a form of social and political action; in a rapidly urbanizing Phnom Penh, heritage buildings like these were being torn down at a rapid pace. Malls and casinos were being built in lieu of public, state-sponsored cultural or green space.

These cinemas were once neutral spaces of levity where urban residents could come together socially for frivolous nights of entertainment—space that was needed in the tumultuous political history of Cambodia. These kinds of spaces were necessary again in the years 2017 and 2018, when Cambodia had entered a politically sensitive period preceding the Cambodian general election (July 29, 2018), which human rights advocates have widely criticized for representing a rapid pivot toward authoritarianism. In November 2017, the Cambodian Supreme Court dissolved the primary opposition party, the Cambodian National Rescue Party, and arrested its president, Kem Sokha, under controversial treason charges. These political events were coupled with increased regulation of media and arrests for oppositional speech on Facebook.

Scholars have theorized that the *ruin* marks destruction: destruction from nature, destruction from people. For Simmel, the ruin represents the force of nature pulling down a structure of human will.[2] Stoler focuses not on the ruin as site of conflict between humans and an outside "nature" but instead on intra-human conflict: precipitating and heralding over the ruin is violence and domination. Ruins show us how past violence continues to persist in people's lives—through the ruined landscapes they live in.[3] As sites of past violence, ruins and rubble are inhabited by ghosts. Sometimes we choose to ignore those who once inhabited ruins, but these ruins do not stay static nor empty. For Derrida, a "specter" is "neither soul nor body, and both one and the other."[4] Like ruins, these ghosts can live in multiple times at once. These specters call our (present) attention to unmarked past injustices.[5] We can also conjure them, perhaps at will. Sometimes they come to the rescue in dire situations.

Gordon writes of *hauntings* that call for present attention to unresolved social ills emerging from historical violence. For Gordon, "the ghostly matter does not just appear and recede as a supplemental effect of real 'objective material conditions' . . . it looks like a *structure of feeling.*"[6] Gordon's analysis of this "structure of feeling," a complex affective matrix, taps into the 1950s Marxist cultural thought of Raymond Williams as well as contemporary affect theory of scholars like Berlant, who chronicles the dramas and adjustments of post–World War II American dreams of the "good life."[7]

This complex affective matrix, in part, defines the media ruin. On the one hand, the destruction of cinema forces us to recognize and mourn death. On the other hand, the ruin gives us space for imagination of happy times. Memories of the movies involve friendship, snacks, laughter, frivolity, romance, and escape. These positive affect–laden associations matter to the way we interpret the media ruin: they represent onetime spaces of fun, entertainment, and sociality.

This contradictory and ambiguous affect also comes to matter in the "death" to media itself. With the death of cinema comes the death of the movie—in other words, the death of representation and fantasy, the death of memories and their interpretations on screen, in soundtracks, and in the hearts and minds of viewers. The ceiling crack of this cinema lets in a stream of light through dusty air, bringing to mind the stream of light of film. Movies themselves act in ways like ghosts. A movie envisions, it pictures a fantasy, and it conjures up magic. As Kittler says, film lets "immortals exist again."[8] The destruction of cinemas then represents the death of movies as ghosts, as memories.[9] Media ruins summon memories of the experience of media viewing and also the memories that the media themselves captured. The ruin is thus fractal, looping into the past. We experience loss in our own minds and hearts, and also through the loss of what once was pictured as pastness on the screen.

This complex and sometimes contradictory set of affects around death, ghosts, happy memories, and intergenerational connections all motivated Roung Kon's project. For Roung Kon, the cinemas were creepy and elicited fear, but they also were important markers of past happy times, of fun, entertainment, and cultural outputs of which they were proud. Roung Kon did their work of surveying, documenting, and touring cinemas as a way to call for more of what they wanted in Cambodia, for their peers, for the "next generation." They wanted more public arts access, better urban planning, and more sophisticated architectural sensibility.

Returning to the cinema tour with which I opened the book, Kagna led our tour group after visiting Cinéstar to twelve of the most famous pre-1975 cinemas, all within a half-mile radius. All of these cinemas were located in the historical northern part of Phnom Penh, rather than the western and southern parts that have been built since the end of the war. None of the cinemas was still running as a cinema, though they were in various states of demolition and required various levels of historical imagination. Phnom Penh streets are often a compelling juxtaposition of grit and new development. The remains of heritage buildings like these are often tucked discretely into recently built offices of concrete. Before the tour, I already felt familiar with the streets of Phnom Penh after years of conducting research there and traveling by foot, tuk tuk, and motorbike. The magic of the cinema tour was that it illuminated anonymous, mundane, and crumbling buildings as significant historical sites. Kagna provided the historical contextualization and imagination to help me envision a cinema at its prime, when people poured in for their evening entertainment.

Following our visit to Cinéstar, we walked to Eden Cinema, a few blocks away, which had been converted into a bank. We looked at the building from the sidewalk across the street. Apart from its art deco frame, it was barely recognizable as an old cinema. The stately large, square structure acted as a frame for a new kind of modernity, with technologies inside mediating financial lives. Kagna told us that Eden once mostly played foreign, particularly French, films, located so close to the old palace and the old French quarter.

We walked a few blocks further north in the city, moving parallel to the brown and smooth Tonle Sap river, a block to our east. We traveled toward the recently restored and now corn-colored yellow colonial post office. All of this walking in Phnom Penh felt unusual and I was afraid someone in our group would be hit by a new "Indian-style" metered tuk tuk whizzing by. We stopped at the former site of the Troung Kok cinema. Kagna talked to a family selling drinks outside their townhouse. We entered the bottom floor entryway of their house. We moved through the bottom floor of their home and exited through their back door. We found ourselves in the middle of a cluster of townhouses built at chaotic angles, leaving an odd bit of empty public space with a gray concrete column, marked by bits of colorful plastic. We huddled underneath a stairway while Kagna explained that inside these homes was a cinema that was built in the 1930s called Troung Kok. The column was an original beam from the cinema. During the 1960s, it became known for showing pornographic films. Hearing this history, I was perplexed in trying to imagine what this cinema could have once looked and felt like. The families who lived in the space went about their business cooking breakfast. As we left, Kagna tipped the drinks seller.

After emerging from the townhouse, we walked two blocks into the city, away from the river into webs of market stalls selling vegetables, flowers, and fish. Crossing Street 13 (parallel to the river), the market maze transitioned to blocks that featured bars with flashing red and black signs and bunk-bed hostels for teenage European backpackers. This area of riverside was, since the entry of the UN in the 1990s, known for sex tourism and cheap outdoor bars selling $1 mugs of watery Cambodian lager. The sidewalks were broken up or nonexistent in these blocks, and I worried again about oncoming tuk tuk and motorbike traffic. With the morning progressing into midday, the sun became hot and I started sweating.

We finally reached the site of the former Kim Son cinema. Instead of a cinema per se, we saw the concrete arch that once led up to the cinema. The carved Khmer script "Kim Son" curved with the arch. Walking under the arch led into a series of alleys where we found small tin-roofed homes and

Figure 4.4
Troung Kok Cinema column

Figure 4.5
Kim Son arch

street vendors camped around colonial-era townhouses. Looking above on this square block from Google Maps, you could see more than thirty little squares and rectangles of homes and shops, not divided by roads or paths, nor in right angles to each other. Now, Kagna explained, this whole block is known as the Kim Son area, even though none of the original cinema building remained.

We walked back down south, toward Street 130. One of the last cinemas on the tour was called the Hemakcheat. One of the most famous cinemas during the 1960s, this building took up nearly a full city block. The light blue writing—Hemakcheat—was still visible from the street, in large rounded Khmer script and smaller Roman script directly below. Kagna explained that the cinema has been a slum since 1979. We walked into the alley directly beside the building; I looked up and saw, for four floors above ground level, people living in open-air spaces. There were clothes hanging from hand railings all the way up the four floors of the building.

What does it mean that this slum was once a cinema, and one that Kagna felt so attached to showing us? These media ruins here bring to mind the violence of their origins and existence: colonial and Cold War roots, national war, and genocide. The poverty of the slums was a new form of inequality and urban violence. Families repurposed these media ruins for ways that were conducive to life, and their survival itself reflected a kind of mundane resistance.

This tour is indicative of the complicated ways that Cambodian young people in 2017 remembered the contested moment of modernity before the war, of which cinemas and movies were important symbols. Cinemas were central parts of a film infrastructure tied to French colonial, American neocolonial, and Cambodian elite nationalist action. As discussed in the first chapter, the Sangkum Reastr Niyum period had unsupportable levels of inequality and gross geopolitical conflict. Because of the violence of the ensuing period (the war and genocide), these ambiguities are often overshadowed by mourning the Khmer Rouge period, including the loss of artists' lives during the genocide. The ways that independent artists coopted these infrastructures for important cultural outputs are more frequently remembered.[10] Media ruins display the ways that the Khmer Rouge put an end to early postcolonial modernist dreams. There is a paradox inherent to these contested legacies of the Cambodian modernist movement being wrapped up in media ruins, which have become sites of urban destruction and poverty.

Roung Kon was led by three architects: Kagna, Daro, and Yury, all in their early to mid-twenties. Kagna finished her undergraduate degree in architecture from the Royal University of Fine Arts (RUFA) in 2017; Daro finished his degree in architecture from a private university in 2017; and Yury graduated from RUFA in 2018. They met when they were all in school while working on a project together about New Khmer Architecture. They did

MEDIA RUINS

Figure 4.6
Hemakcheat

some specific work about Van Molyvann, the most famous Cambodian modernist architect, in collaboration with another research group called the Van Molyvann Project.[11] They started to learn how to do research and became passionate about heritage buildings in Phnom Penh. As a part of this project, they learned about Capital Cinema, designed by Van Molyvann in 1965. They were all distraught to see the cinema being torn down in mid-2017 to make room for new urban buildings, and they told me that they "wanted to take action."

The cinemas were more than just places to watch movies; they were places for hanging out and being together. Yury told me: "It is amazing to think of all the people wearing beautiful clothes, looking sophisticated at the cinema in the previous time. This is about our city, our lifestyle, our culture, our identity." This project helped the team have more dialogue with older people in their lives. Yury said this project allowed her to talk with her mom, who told her about her experiences at the cinema earlier in her life.

The mystery of the cinemas drew them in as well. "Most of the cinema is gone but we can still see some structure that still exists and we want to know more about it," Daro said. They were frustrated with their peers who don't seem to be as interested as they are in these old cinemas. They blame the cinemas' decline on their peers' failure to patronize them. Now young people, Kagna explained, "want to go to new malls for watching movies." It was easier to park and find food around the malls. They also had "these new phones" and other new technologies for watching films.

The last of the heritage cinemas running in Phnom Penh, Lux Cinema, closed for movie-watching in January 2017. Yury, Ro, and Kagna were sad it closed and blame its closing on their peers not wanting to go see movies there. It was a scary place, though, and even the Roung Kon team avoided the cinema at night.

Kagna: I'm not too surprised it closed, I think it was creepy and haunted.

Ro: It is an old building—an old building that went through the Khmer Rouge. After the Khmer Rouge, no people wanted to go inside.

Yury: There are some ghosts inside.

Kagna: It can be really scary!

Yury: We went there before but we never saw them. We just heard from other people.

Kagna: But also—at night, it's not good!

Yury: But we interviewed the wife of the owner of the cinema and asked about the ghosts. She said that some people see them.

Kagna: I tried to bring my friend and he said, "No, I don't want to go inside; it's too scary at nighttime."

Kagna, Ro and Yury's fears about ghosts seemed to me somewhat tongue-in-cheek; they laughed and talked about ghosts in a halfway joking manner, but I sensed a real sense of fear behind their smiles. The cinema is often remembered not only as a haunted place but also as a place of refuge for urban Cambodians. An older Cambodian friend always laughed when he talked about the old cinemas, bringing a dark humor to his memory. During the civil war of 1970–1975, he told me, people would still go to the cinemas even when Phnom Penh was being bombed. "The cinemas were fun! They were entertaining! It was the only entertainment in the city then." When the cinemas became a target for Khmer Rouge bombs and people started to get killed, people kept going. Until April 1975, the cinemas were open and new movies would sell out. In the Khmer Rouge period, buildings like these were repurposed or closed.

Throughout the period that Roung Kon was working (roughly 2017–2018, though their project continued part-time for some years following), urban development in Phnom Penh happened unbelievably quickly. Heritage buildings of all kinds were being torn down to support the huge number of new high-rises in the city. Kagna once told me that she thinks that—from an architectural perspective—2017–2018 was almost as bad as the Khmer Rouge period since buildings were being torn down as rapidly. By late 2018, within the two and a half years that Roung Kon had been running, nearly two-thirds of the cinemas that they had first surveyed and put in the tour had been torn down. There were rumored plans to tear down Lux Cinema.

During this period, according to the United Nations Development Program, Cambodia was in its fifth year of 7 percent GDP growth.[12] The cityscape itself was a sea of cranes, construction sites, and demolition. My apartment was on the seventh floor of a new apartment building in the Psar Toul Tom Pong neighborhood (Russian Market), a rapidly gentrifying neighborhood south of central Phnom Penh. I had a gym on the tenth floor of my apartment building and would spend some evenings walking on the treadmill, watching the orange pollution-filled sunsets over the city. I would count the number of visible new building construction projects—wrapped in a distinctive green plastic—into the tens and then the twenties. There were three new high-rises built in a two-block radius of my apartment within the twenty months I lived there. From my seventh-floor balcony, I would wave to the construction workers and worry about their safety as they climbed without ropes high in the sky. I started to track time by the height and completion of buildings surrounding my balcony.

Directly below my balcony, though, was a single-story noodle factory. I could also watch every morning at 6 am as the noodles would be hung to dry outdoors, on horizontal rods. Across from the noodle factory was a traditional two-story house. I could watch the family that lived there play in their front yard. Throughout the course of my residence, the family was potty-training their toddler; her grandma would chase her around the yard with a portable potty. I wondered what this family, and what these noodle makers, thought about my apartment building and the high-rises going up rapidly around them.

Political change accompanied this huge and uneven economic growth. Human rights advocates have widely criticized the 2018 Cambodian general election as a strong pivot to illiberal democracy with the ruling Cambodian People's Party (CPP) dissolving the opposition Cambodian National Rescue Party (CNRP).[13] In the preceding year, the governing party disbanded media independence and the primary opposition party. On September 3, 2017, Kem Sokha, the president of the CNRP, was arrested for treason; the government cited a speech Sokha gave in Australia that referenced a partnership with the United States as grounds for his arrest.[14] On November 16, 2017, the Cambodian Supreme Court dissolved the CNRP.[15] Finally, in July 2018, the Cambodian general election took place with the CPP declaring victory.[16]

A media crackdown preceded this political crackdown. At the end of August 2017, thirty-two independent (nonstate) radio stations were forced to close or stop broadcasts of Voice of America and Radio Free Asia (seen as oppositional radio).[17] In the following weeks, the government closed the *Cambodia Daily*, a prominent critical English-language newspaper based in Phnom Penh, after a tax dispute.[18] The internet came under more state sector scrutiny after these events in the traditional media sector. On November 14, 2017, Oun Chhin and Yeang Sothearin were arrested for operating

Radio Free Asia on Facebook and operating in a Phnom Penh office (they were released on bail nine months after their arrest).[19] Before the end of 2017, seven people were arrested for statements against the prime minister on Facebook, by far the most popular online platform in the country.[20] In May 2018, the Ministry of Interior, the Ministry of Posts and Telecommunications, and the Ministry of Information declared they could monitor and control the internet and make arrests based on online activity.[21] The week before the July election, access to independent news websites was cut off in Cambodia.[22]

Facebook and smartphones have had an important impact on the political sphere in Cambodia. As with the introduction of Radio UNTAC, the rapid increase in mobile-enabled internet connectivity in Cambodia has challenged historically state-controlled media channels such as radio and TV. Scholars have argued that rural and urban people, traditionally disenfranchised from political participation, became more politically active during the run-up to the 2013 national election, a hotly contested election between the ruling CPP and the opposition CNRP, because of increased access to discursive discourse on social media. There was a momentary feeling of internet freedom (though many fewer internet users), and people could criticize the government and call for change on Facebook.[23] Political participation on social media did not often take the form of overt calls for regime change, but rather calls for shifts in "everyday" issues.[24] Hughes and Eng argue that the rural poor in this period were far more likely to "share" and "like" newly available political and oppositional information, but it was uncommon for this demographic to write their own political opinions via comments and posts. They call this form of political resistance a "quiet encroachment on real-world distributions of power."[25]

The 2017 media shutdowns and the government's commencement of arrests for critical Facebook activity shifted this temporary moment of some internet freedom. Because of these new information rules, in the run-up to the 2018 election, many users felt much more constrained in their use of the tool than they had in the 2013 election, back to a sort of status quo of restricted media freedoms that predates UNTAC. Cambodians across demographics reported avoiding posting politically sensitive material on Facebook. Beban and colleagues empirically show that, for opposition-leaning journalists, social media once felt like a space of "both emancipatory and authoritarian potential."[26] They argue that "the Cambodian state's crackdown on media is part of an ongoing transformation of authoritarian populism that has reduced the space for rural collective action" and that "the crackdown signals the loss of a more emancipatory, democratic imaginary."[27] In addition to new information rules and controls, pro-government political activity on Facebook is encouraged by the state and is a part of many government workers' jobs. Government workers feel social pressure to use Facebook to support the ruling party.

The city—and the Cambodian internet—became a scarier place through the course of the 2017 government tightening of the online sphere. Everybody that I spoke to in Phnom Penh had become afraid of the possibilities of sharing, particularly on Facebook, because of fear of political sensitivities and unknown negative consequences. This fear took on tangible qualities around the election period. In casual conversations, some friends would brush off the new media censorship and the authoritarian crackdown as just more of the same in Cambodian politics. Others would quietly express frustration about the transition between the 2013 and the 2018 elections. One participant (a twenty-one-year-old man from Phnom Penh) told me, "Before we shared, now I rarely post. Before I didn't know, now I know better. Before I always shared, now I never share. If we put up sensitive information or 'hot news,' this can be a problem, or a challenge on Facebook. You cannot post, you cannot like what you want. We are scared. You cannot do what you want." These economic, architectural, and political events in Phnom Penh have urged my interlocuters to establish private and physical spaces. "We need places where people can talk confidentially and with trust," another friend told me.

Media ruins do not stay in the past, but occupy multiple temporalities (past, present, future) through ghosts. Stoler explains that "to think with ruins of empire is to emphasize less the artifacts of empire as dead matter or remnants of a defunct regime than to attend to their reappropriations and strategic and active positioning within the politics of the present." Stoler urges us to see the ways that ruins move us also *forward* in time: "Ruins draw on residual pasts to make claims on futures." Today people living amid ruins use these spaces of past injustice to make a livelihood, have a home and a family, construct a sense of well-being, and call for change. Building on Stoler's emphasis on the forward action of the ruin, Gordillo introduces *rubble* as "a set of objects and a concept" that ground his ethnographic analysis of sites of destruction in the Chaco region of Argentina.[28] He points out, in various empirical examples, that "rubble and insurrections" are intimately connected.[29] Gordillo also builds on Walter Benjamin's idea that destruction can "contribute to a collective awakening from the nightmare of the bourgeois dream world."[30] Our protests can make full use of the artifacts at hand; sites of the past, casually yet confidently decaying, work as painful or joyful reminders that things do not always have to remain as they lay. Gordon too paints a picture of the ways that "ghosts of past violence continue to keep urban memory of such acts alive in the present."[31] Hauntings tie the past to present political action, leading us to the futures that we want. They signify future possibility and "ethereal intervention" into what might come to be.

Ghosts are both a metaphor and a matter of belief in media ruins. According to Southeast Asian cosmology, ghosts routinely inhabit abandoned and destroyed buildings across Southeast Asia.[32] These spirits include the Theravada Buddhist and animist "people-land" spirits *naek ta* in Cambodia.[33] When abandoned buildings become sites of urban transformation and development, their ghosts and enchantments continue to mediate social action into the future. Schwenkel demonstrates how what she calls "haunted infrastructure," simultaneous sites of pastness and urban development, become sites of contestation of state power during periods of destruction and (re)construction.[34]

Likewise, media ruins act as sites for contestation of power and the construction of futures. These ruins suggest the spaces for the arts that could exist again. Roung Kon's work does not just mourn the death of cinema; it also indicates the possibilities of restituting a media space. The ruins of cinemas map the intersections between Phnom Penh's neoliberal urban development and the rise of information control. Through documentation and surveying, Roung Kon takes control over their imagination of the space that once surrounded the imagination of film. Their work therefore gives insight into the power of infrastructural restitution as resistance to information control and neoliberal urban development.

Documenting the cinemas takes careful research and patience. Roung Kon's project was a good way for the team to test out their surveying and research skills and to learn new approaches. They interviewed older people who were alive in the 1960s and early 1970s in Phnom Penh about their memories of the cinemas. They also traveled to provinces to interview people about heritage cinemas, including to the capitals of Kampong Cham,

Battambang, Kratie, and Kampot provinces. They completed document review online in institutional databases, Facebook, and in the few paper archives in Phnom Penh.[35] They made maps of the cinemas they found in Phnom Penh, Kampong Cham, and Battambang, available on Google, linked from their website and Facebook page, with pictures of what the cinemas look like now (often refurbished into new spaces) and what they looked like before the Khmer Rouge.

Some of their most important research outputs are surveys of cinemas. The goal of the surveys was to document the architectural layout of the cinemas in their heyday, before walls were torn down or added for repurposed buildings. They found columns and important architectural features that gave form to the original structures. They contextualized the survey with historical documents about the cinemas in archives, though these can be hard to find. They took photos of what remained, including 360s, drew the cinemas in AutoCAD, and then rendered the cinemas in SketchUp. They referenced their photos when they had questions about the cinemas' outlines. Some of these surveys were the basis of wooden models. Their ultimate goal was to create an architectural archive of these surveys for the "new generation," so they could see the buildings after demolition. New architects can thus learn from their work and the historical urbanism of the city.

I accompanied Ro to the Chenla movie theater for a surveying trip. We were joined by Pheary, a younger architecture student who volunteered for the project. Ro told me that this is a complicated building—it isn't square. He said, "Now buildings have right angles but in the 1960s they had strange angles." Chenla is by far the most well-preserved heritage cinema in Phnom Penh; the state and private companies gave some funding after the war to renovate this cinema into a theater in the 1980s and 1990s.

Ro and Pheary took measurements of the length of each wall. Ro worked in the front of the theater by himself, while Pheary worked in the back. Ro was able to borrow a laser measurement tool from a friend. He said, "I didn't learn the process of using a laser measurement when I was in school, just picked it up after and I borrowed it from a friend who was doing freelance work."[36] Pheary worked in the back with a measuring tape. The measuring tape was a little too short for one of the walls and she had to mark the spot where it ended and start measuring again. She mapped the walls on paper and decided which segments to measure, writing the measurements one

Figure 4.7
Chenla Theater

by one on her paper map. She told me, "I start with measuring the length of every wall, then I mirror the other side for symmetry." She and Ro each entered their measurements into AutoCAD.[37]

This work took patience and was time-consuming. Pheary and Ro had started the previous weekend. Since they are also working full-time during the week, they came to the site only on weekends. The government ministry that owns the building had given them a permission letter to do this work for three consecutive weekends. They were able to recruit some other volunteers to help, and five other architecture students had worked with them the previous weekend. Everyone was learning proper survey techniques through the project; they all knew how to survey from school, but slightly differently, so they collaboratively learned and taught each other.

Pheary at one point was confused when she tried to figure out how to measure the complicated angles by a doorway. Pheary at first looked frustrated, but Ro (who was a few years older and more experienced) jumped to help her as soon she went to the front of the theater to ask him a question. After a few minutes working together on her section, they realized there was an issue with the angle ("we were missing a thickness near the door").

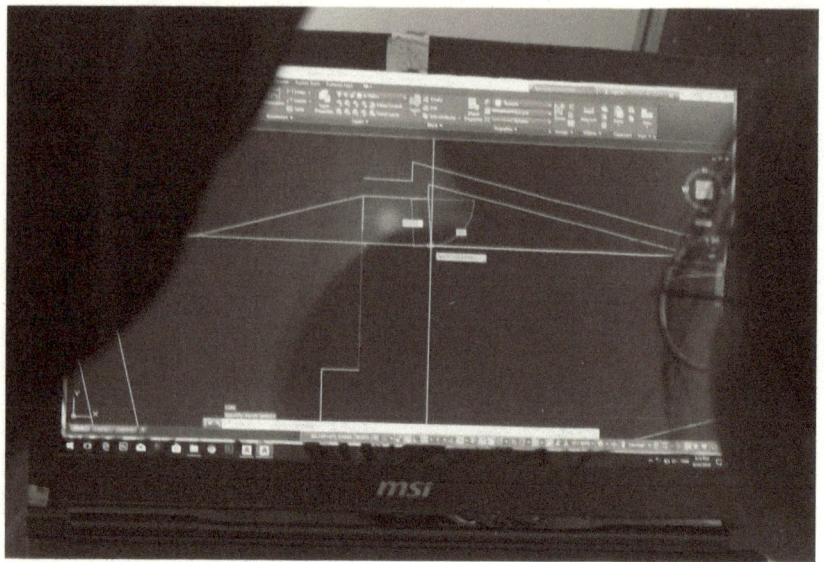

Figure 4.8
Surveying at Chenla

Ro drew the angle on paper. He used the laser for more precise measurements. They worked together and entered the figures in AutoCAD.

Keeping up with all of this labor wore out some of the team. Over the course of two years, Roung Kon continued part-time. Ro, Yury, and Kagna had some trouble getting consistent sources of funding, so they all got other jobs on top of this project. Kagna started to give two tours a day at Khmer Architecture tours and taught interior design once a week at Limkokwing.[38] Cobbling together work like this gave her better overall pay and hours than working at a private architecture firm.[39] Ro, though, decided to start working full-time at an architecture firm. Yury started working part-time at an art gallery and had an independent art show at a cafe in 2018. Despite other projects, the three continued to work on the Roung Kon project together in their free time.

Yury told me once that they were all so busy trying to make enough money to survive that it became difficult to find enough time to meet. They were not able to keep their website up to date. She had not been feeling well, either, and she had been feeling very stressed, leading to some uncomfortable health problems. Kagna also took time off for stress-related illness. Sometimes the group had trouble getting access to the inside of the cinemas; they needed to get a special letter from an NGO and ask for special access permission. They sold branded journals and postcards as a way to make some extra money, but it was not enough.

Roung Kon had important successes, too. They were able to exhibit their results in various galleries, cafes, and special events in Phnom Penh. For example, at an exhibition at the "City for All" conference in November 2017, they showed photographs of some of their favorite cinemas, photographs of themselves touring the cinemas, tiles from the original cinemas, and old advertisements for films in *Réalités* magazine from the 1960s. Kagna and Ro were both there and told me how happy they were to get attention for the project.

This project at the sites of media ruins is intricately tied to building new futures. At a time when the internet was becoming less safe and spaces for free expression were closing down, Roung Kon did the patient work of documentation of a time when arts flourished and the government and broader society widely cultivated public space for entertainment and the arts. The Roung Kon team worked across temporalities: they celebrated the achievements of past architects and filmmakers; they rejected the urban destruction happening in 2017–2018; and they gave their own resources to build a future of cultural promise. This project of future-building went beyond just surviving amid "imperial debris." They worked together, sacrificed time, and integrated diverse resources. What is more effectively optimistic and fruitful, in a time of political and architectural restriction, than patient, creative, and collaborative work?

One reason I came to know the Roung Kon team was because we worked together at Kon Len Knyom (meaning "my place"), an independent art space dedicated to building art audiences and networking run by Meta Moeung. Kon Len Knyom mirrored Roung Kon's political action in its creation of a physical space and in-person network as a form of "invisible infrastructure" for community resilience and the encouragement of a new art sector driven by Cambodian young people.

On her website, Meta described Kon Len Knyom as "a place for art students, independent artists, curators, and researchers looking for an inspiring space to network and work on their individual projects . . . We aim to connect, share ideas, and facilitate artistic knowledge between visiting artists, curators and researchers." Kon Len Knyom was founded in February

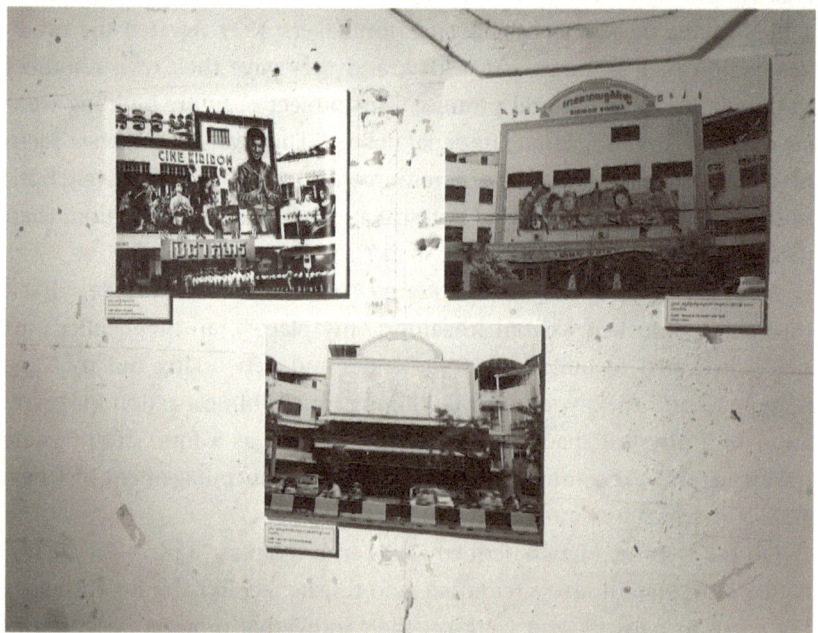

Figures 4.9 and 4.10
Roung Kon exhibition at "City for All" conference (at the "Mansion" exhibition space, Phnom Penh, November 2017)

2017 and was located near the Toul Sleng Genocide Museum. Meta, ten years older than the Roung Kon team, started the space using her own money from independent arts management and administration positions, including the income she generated as the executive assistant for an internationally recognized Cambodian visual artist two days a week. Roung Kon was the first group that she incubated, and she gave them free space to work, meet, and hang out. At the same time, Meta also gave me space to work on my research.

The Kon Len Knyom space was in a traditional townhouse that was hard to find for new visitors. Down an alley and to the right, the space was obscured behind a metal grate. The entrance held a few outdoor tables, protected from rain by an overhanging second floor. Upstairs were two small rooms with desks, rarely used. The main room, downstairs, was an open space with cushions and short collapsible tables for computers, which could be rearranged depending on who was working in the space and where they wanted to work. Along the edges of the room were bookshelves with an eclectic library of Khmer and English language books about art, architecture, Cambodia, and Southeast Asia. The back of the room had a small kitchen, from which Meta served guests Vietnamese, Cambodian, and Chinese teas. Meta picked up street food from nearby stalls, which she brought back to the space to eat with the others. She served the food for everyone on the beautiful plates that she had collected, many from a Japanese secondhand store. Meta often decorated the space with fresh flowers. Students and researchers helped with upkeep and maintenance, including painting and taking care of the plants. One day in October 2017, we together painted the outside gate red.

Meta told me once why she built Kon Len Knyom:

> It is for the future. It is about people who use the space. . . . I want to create what I call an *invisible infrastructure*—you can know each other from this space. . . . In this invisible infrastructure, everyone is respecting each other, everyone is responsible for themselves, they know each other through the space, it allows people to have discussions, it allows people to be free without framing. You just feel like home. When you are here, you can talk, people will come, students will come. . . . When you are independent, your brain gets to be independent. [Cambodian] young people need inspiration . . . they don't have a lot of access to this sort of thing.

Kon Leng Knyom also acted as Meta's unofficial gallery. She attended nearly every art event in Phnom Penh's burgeoning art scene, spread out between only a handful of spaces, the most important (at the time) being Java Cafe, Sa Sa Art Projects, and Sa Sa Bassac.[40] She beautifully decorated Kon Leng Knyom with the art that she has collected through the years at these events, some of which have been gifted to her for her generosity and expertise from major artists in the city.

Kon Leng Knyom also hosted events, usually about three evenings a month. These included presentations about research, arts, and student projects from a mix of Cambodian and international artists and researchers. Along with events at other spaces (such as Sa Sa Art Projects and Java Cafe), these events were important opportunities for researchers and artists to come together, meet, discuss collaboration, and get feedback on new projects, part of the "invisible infrastructure" that Meta has sought to build.

Yury, Daro, and Kagna's work at Roung Kon was possible with the support of Meta: her encouragement and space. They had no places to work outside their family's homes and school. As volunteer workers, they had no way to pay for a private office. In a time of information control, media spaces like the cinemas whose memory Roung Kon sought to preserve again became critical spaces of sociality, creativity, and personal expression. Kon Leng Knyom was not showing films the way cinemas once were, but it filled some of the same basic needs of the members of Roung Kon. In the context of Phnom Penh in 2017–2018, the construction and maintenance of this space of trust, community, and artistic expression were an important form of self-protection and radical action. Brick and mortar spaces of trust became only more important as censorship and control intensified.

The work of Kon Leng Knyom and Roung Kon was intimately related and represented a trend I noticed across Cambodian young media creation spaces of friendship, volunteer work, collaboration, and support. These projects were acts of care and attention to detail. Consideration to small encounters made life joyful and community-oriented. These spaces gave room for free thought, independent ideas, and trust in an otherwise difficult time when media and free expression were waning, and the internet no longer was a safe place for open dialogue.

The work of Roung Kon, Kon Leng Knyom, and the concept of *media ruins* help us to contribute back theoretically to the literature on ruins and the ghostly by querying the special relationship between media, memory, and space in a postcolonial and post-conflict environment like Phnom Penh. Media ruins, like ghosts, take on contradictory "structures of feeling," from mourning death to celebrating past arts accomplishments, from memories of bombs to memories of sophisticated evenings of entertainment. In Phnom Penh and Battambang, cinema spaces that once hosted artistic expression and acted as happy places of sociality were being torn down, during a moment when the government chilled other kinds of free expression. Cinemas were sites of urban destruction and construction, part of a trend of urban gentrification that exacerbated social inequality. The ways that media ruins came to embody these complex feelings resemble the ways that movies themselves elicit multiple and ambivalent affects. We choose to go to the movies, sometimes to see painful past memories and sometimes to see love stories or comedies. Movies capture life, which, when lived fully, contain multitudes of affect.

The future-building action of Roung Kon moves beyond just making meaning out of the past. Their infrastructural restitution—their careful surveying and displaying of heritage movie houses—subtly protests against recent urban changes and new forms of internet control. Media ruins as mediated by Roung Kon need to be documented because they are places not only of pastness but also of promise—promise of collaboration, trust, artistic expression, and entertainment. In the context of Phnom Penh, this infrastructural restitution is a powerful but subtle call for political and social change. Their hosting in Kon Leng Knyom reminds us that building and maintaining brick and mortar spaces of independent arts and creative thought are radical actions. They call for a Phnom Penh with public arts, green space, and better urban planning. As Stoler notes about imperial debris, ruins are sites of change and building futures. As Gordon argues, ghosts remind us of past wrongs in order to call for change. Roung Kon's work demonstrates that media ruins too act as sites of inspiration for political and social future-building.

Conceptualizing Roung Kon's work through infrastructural restitution gives insight into how often unrecognized labor of repair and reconstruction plays a role in social and political action. Yet I want to note here that their infrastructural restitution is not neutral. The specificities of a place—its historical conflicts and contemporary political workings—give the work of restitution particular kinds of meaning, meanings that are actively contested between varying agents. The meanings and interpretations that Roung Kon subscribes to media ruins and their restitution are only one kind of possible meaning. Another way that media ruins here take on a complex set of associated affects includes their association with a history of colonialism, a contested modernism, and domestic inequality. Those who were on the other side of the civil war that broke out at the end of the Sangkum Reastr Niyum period might challenge Roung Kon's telling of history. Because the Khmer Rouge period was so painful and the regime ultimately lost, the political stances of the Khmer Rouge, which sometimes encompass the calling-out of the destructive forces of imperialism, colonialism, and elitism, have become to some extent marginalized within Cambodia.

5
DISINTEGRATION NOISE: PREAH SORYA'S FILM RECONSTRUCTION AND HEALING THE "WAY OF THE HEART" (*PHLAUV CHETT*)

> I watch these original films many times because I want to remember (*chong cham*). I watch them again and again because they make me feel (*mean aaram*). When they cry, I cry . . . when they are happy, I am happy (*kay sabbaay, khnom sabbaay dei*).
> —Vandy, Preah Sorya member

On a homemade rooftop cinema in Phnom Penh in September 2017, I joined a group of young film collectors to watch a grainy and soundless but richly colored shot of two actors, Sam Oeun and Kim Nova, rowing around a lake that once existed in the city. The scene cuts and repeats, the actors making slightly different gestures, indicating the director changed his instructions for this second take. This cut comes from *Golden Violence*, a film from 1967–1968—the height of Cambodia's film industry. The two stars were killed in the 1975–1979 Khmer Rouge genocide, and most of their films were destroyed during the regime. The group that has hosted me calls themselves Preah Sorya (Sun God) and they found this unedited version of this film on a reel with three other films in a collector's shop in Kampong Cham province, then digitized it at the Thai Film Archive in Bangkok. No full version exists. The group has traveled extensively through Cambodia and internationally to collect films like this one from before the Khmer Rouge regime, which they then repair, digitize, and disseminate through public film screenings and on Facebook.

This chapter explores how infrastructural restitution builds community through loss, realized materially as glitches. The films that these young people collect and document have a beautiful and romantic aesthetic of breakdown. What I call the films' *disintegration noise*—resulting from physical breakdown, the transfer of film from one format to the other and the marks of reproduction of image, partial versions, and unedited scenes—makes the past visibly present while watching them today. I argue that this disintegration noise, rather than detracting from the films' value, emphasizes the film's status as old and original, and makes them more valuable for their collectors since historical documentation about and images of the past are scarce and deteriorating in this context. The noise also reminds the viewer of something of Cambodian history, both the peaceful and playful realities on screen as well as the period of war that occurred between the time of the production of the films and today.

Disintegration noise opens up the films to new creative possibilities for commemoration of the past. Preah Sorya's project documents and celebrates artists and a way of life that was lost during the Khmer Rouge/war period. Their project, though rooted in the past, helps them experience emotional healing and dream toward an emotionally healed future for themselves and their peers. They embrace the imperfections of the media they find and use them as an opportunity for producing new outputs interweaving the films, music, and live reenactment. They do so by putting to work new media tools as well as transnational networks, in-person film screenings, oral history interviews, and extensive provincial fieldwork.

My participants tell me that they use their film recovery and presentations as a way to heal the *banhhaa phlauv chett*—literally translated as *problems of the way of the heart*—a term used to describe the deep sadness and emotional and mental difficulty emerging from the experience of painful events and loss (often referring to but not limited to events of the Khmer Rouge period). Watching old films helps to heal both their own *banhhaa phlauv chett* (though they did not live through the Khmer Rouge themselves) as well as those of their parents' generation who more directly experienced the war period. The ways that Preah Sorya uses collected media to move past Cambodia's violent past demonstrates a new dimension in the relationship between media and memory.

As described in the first chapter, in the late 1960s and early 1970s, moviegoing was a popular activity in Cambodia, and more Cambodian films were made in this period than at any time before or since.[1] Because of the dynamics explained in earlier chapters, it wasn't until the early 1990s that the search for old films could begin openly and in earnest. In 2010, there were only thirty-three known films, mostly sold on pirated videocassette versions.[2] As of this writing, 100 of the approximately 500 films made during this period have been found, many of them in partial format, and they are not collected in a centralized institutional archive.[3] Restored films have been found in haphazard locations such as private homes, spared cinemas, or in foreign countries, as many had been taken abroad during the war period. The lack of institutional attention to and funding for the search for these films has led to the need for volunteer groups such as Preah Sorya, the Cambodian youth group researching and collecting old films, whose activities I describe in this chapter. The findings in this chapter are based on semi-structured interviews, participant observation, and informal conversations with the group founders, audience members, and other Cambodian media creators and reconstructors.

Cambodia's relation to media has been changing dramatically, and the Cambodian population has begun to rapidly adopt digital tools. Media production has shifted from predominantly analog and film-based to various digital forms, and distribution has shifted from controlled spaces such as private homes and the cinema to the internet, primarily accessed on mobile phones. The contemporary Khmer-language film market, for both commercial and arthouse releases, is small, but many of my participants feel optimistic about its growth potential. Many Khmer films are first released in new commercial theaters and then on the internet. These changes in materiality and distribution have important consequences, such as increased viewership and expanded notions of connections to global media production. This change in the media landscape in Cambodia also shifts the way that restoring media is done and thought of as a project of national recovery.

Muouy Meun Allay (in Khmer, "Ten Thousand Regrets," and the name of a popular melancholic song and film from the 1960s) and was the name of a three-day film festival that Preah Sorya organized to screen recovered films from the late 1960s and early 1970s. The festival took place at Chatamouk Theater, a 1960s modernist theater now used for public cultural events located on the Mekong River in central Phnom Penh. As I walked into the theater on the first day of the festival in August 2017, I was greeted by a team of around twenty volunteers, who are mostly students from the universities around Phnom Penh. The group sold tickets for 10,000 riels, or $2.50, per movie and has made and is selling books about the 1960s actors and actresses, bags, and T-shirts for fundraising.

Preah Sorya began a project in 2010 to find and disseminate prewar film; the founders and participants in the group were a mix of university students and recent post-grads aged eighteen to twenty-six years old. Many of them were involved with the project because they wanted to know more about the golden age of cinema in Cambodia. Vandy, one of the leaders of the group, told me he wants to do the work because he finds the films so aesthetically beautiful. He also said it is important for him to keep alive the legacy of the actors who died and thinks it is important that his generation know about their creative legacy.

The first film I watched at the festival was called *Neeung Jew Kreu Fa* (the name of the protagonist), which was made in 1967 and opens with the young daughter of an earnest woman being kidnapped by evil witches as the mother and daughter travel through the forest. The materiality of the film is striking: first made on film in 1968, it was videorecorded using a camcorder and transferred to VHS sometime in the 1990s. The Preah Sorya team bought the film on VHS in 2010 and then digitized it at the Thai Film Archive in Bangkok. Now the remnants of the three media modes are clearly apparent in the viewing of the film. Occasionally cutting out, showing the horizontal stripes of black and white blur of the VHS format, the spots of the film, and the watermark of the digital platform, the translation of this memory from one era to the next was a striking part of the viewing experience.

Virak, one of the founders of Preah Sorya, explained how the group found the films in various formats through extensive digging in old shops in Phnom Penh, across Cambodia and abroad. They first started searching in collection shops in Phnom Penh. They traveled throughout the country to collect materials. They found film reels in Siem Reap, Kampong Cham, and Battambang provinces. Virak then traveled to Thailand, where he was able to find some Khmer films that were brought there and dubbed into Thai in the 1960s and 1970s and which are now stored at the Thai Film Archive. He has subsequently gotten copies and redubbed them into Khmer. They also found photos, posters, and other ephemera from this era through connections to Hong Kong and France. Sometimes for significant sums (thousands of dollars), they bought these films and pay for their repair and digitization at the Thai Film Archive.

The films have noise because they were not stored properly or were moved often during and after the war. Vandy later told me about the origins

Figure 5.1
Still from *Neeung Jew Kreu Fa*, with cassette and digital marks visible

and the significance of the noise in the films they collect. He explained to me, "When we transferred them, we edited the film. We looked at them frame by frame, because some frames had noise and some frames didn't. Some frames were completely destroyed. In the places where the film was completely destroyed, we cut the image out of the film. Sometimes we could use the other side of the film. Sometimes, there was just a dirty mark and we were able to keep it." I asked him what he thought about the noise, and he responded, "Sometimes, in new movies, they want to make noise as a style to make them look old, to make them more interesting. But this isn't good—it isn't real. In these original movies, they have noise because they are really old—and that has value."

After the screening of *Neeung Jew Kreu Fa*, the team hosted an opening reception with nearly every seat of Chaktomuk Theater occupied (it has a capacity of 420). The event opened with thank-you speeches to sponsors, mostly local companies and individuals. Vandy gave the main speech of the night, which climaxed when he held up an old film reel and gave a hearty thank you for the financial support from the sponsors. These films, he explained, were expensive to buy, but he emphasized the group's

Figure 5.2
Vandy with film reel at the front of Chaktomuk during the Muouy Meun Allay event

commitment to research and their experience traveling through the country to collect films and other ephemera. Vandy told me later that their budget is one of their biggest challenges. They have enlisted some sponsors for big events like the film festival; they also used some of their own money for events and finding new material. They asked for donations when they can, especially for special screenings. They also sold T-shirts, bags, and books at large cultural events around the country.

The evening ended when the group introduced three famous artists from the 1960s. I realized through the din of loud clapping the true enthusiasm of the audience for this moment. They introduced each guest with a video montage before they slowly walked on stage. Tap Songva, an old musician, came out first. Then the group introduced Sar Kassora, a former actress who now lives in the United States, and Dy Saveth, the most famous living actress from the 1960s, who moved to France during the war and has subsequently returned to live in Phnom Penh.

On Sunday evening, I returned for the final event of the festival. The event started with Cambodian young people modeling 1960s-style clothes to the background of 1960s songs. We watched, for example, a young man

Figure 5.3
Muouy Meun Allay closing event

and woman gracefully walk around the stage to a 1960s duet of Sinn Sisamouth and Ros Sereysothea (two of the most famous 1960s Cambodian musicians), in formal Western attire. Then we watched four young men in trousers and short sleeve shirts, and four young women in a mix of Khmer traditional silk *sampots* and Western outfits, coming out to dance the Khmer version of "go go" dancing.

After the modeling finished, Vandy introduced the film of the evening. Made in 1962, the film, *Debt*, was famous in Phnom Penh throughout the 1960s. It is a comedy that portrays a cross-section of Cambodian society during the 1960s Sihanouk-dominated era, from Cambodian modern urban young people to their traditional parents, spiritual guides, and rural peasants. It features some of the most famous actors and actresses from the 1960s. Preah Sorya described the film with the following blurb: "*Debt* discusses the problems of Cambodian young people and their families, and shows the sacrifices that young people make for their society. It shows beautiful clips from 1960s Phnom Penh and Bokor Mountain."[4]

The experience of the film showing was aesthetically striking. The group found only a partial version of the film reel and it starts about midway through the length of the film. They digitized the film directly from a film

Figure 5.4
Still from *Debt*

print (rather than from a VHS like *Neeung Jew Kreu Fa*) and the colors are bright and vivid, despite some weathering of the film.

At the beginning of the clip, we watched young people from the city get into a car and drive to Bokor Mountain (a vacation location and former French hill town in Kampot province). They see a guru and ask for a fortune telling; the dubious healer turns a duck into a snake into a rope. The film skips a few sections, then we watch as the car breaks down on the side of the road and the teenagers try to figure out how to fix it. The next scene is back in Phnom Penh and one of the boys starts to court one of the girls. Her father is difficult to please, and we witness a clash of traditional values and modern romance. One of the grandparents becomes sick and the teens help. Then the film abruptly ends without the original conclusion, leaving the audience with a mysterious and beautiful glimpse into 1960s life.

Figure 5.5
Group of young actors playing 1960s roles

After the film finished, the event took an unexplained break. We could see from our seats that actors were moving backstage. After a few minutes of rest and silence, the actors came out again and lip-synced to a 1960s song. Though this act was not in thematic continuation with the film, this reenactment felt as if they were reconstructing the part of the film that was lost. Virak ended the evening after the performance with a deeply emotional speech. He teared up and dedicated the event to the actors who were lost.

Vandy said to me an in an interview, "The films help older people [who lived through the Khmer Rouge] feel that they live in this happy past time [captured in the films] and that they can forget the time of the Khmer Rouge for a short time. The films help them to heal the *banhhaa phlauv chett* for a short time." *Banhhaa phlauv chett* is a way to describe the deep sadness and emotional and mental difficulty often emerging from a painful event. He continues, "One mother who lived through the Khmer Rouge watched an original film yesterday. For her, watching the film made her happy but it made her shocked.[5] Society is progressing and happier, but this film reminds her of being a child. Reminds her of the river, reminds her of the cyclo, reminds her of the marriages, of studying, and the film tickets that they gave away at school. The films for her make her happy, excited, but also almost crying."

Vandy and his friends, however, did not live during the Khmer Rouge period. He told me, "We are not the victims of the Khmer Rouge. But I am deeply regretful—we should not have had the Khmer Rouge regime. They should not have destroyed everything—for people, for society, for the nation. They should not have destroyed the good things. I am so sad, regretful." But for him, too, his *banhhaa phlauv chett* can be solved for a short time by watching the films and having the happy feelings that they elicit.

The final effect of the evening was spectacular. Though these films are what we might consider bad quality under different circumstances, the combination of the screening and the live reenactment was indicative of a high degree of care and appreciation from the Preah Sorya team. They filled in the literal holes of the films with striking live performances, with an almost eerie effect of seeing the past come alive again. By bringing together actors who survived since that time period, they also acknowledged through this performance the ways time has passed.

Many years of intense national conflict have left deep emotional imprints on Cambodia and Cambodians, whether or not they have lived through the crises themselves. There is discussion and debate about how useful the Western medicalized notion of post-traumatic stress disorder (PTSD) is to describe the experiences of survivors of the Khmer Rouge and their families. Some researchers cite a high incidence (though still, many claim, largely under-diagnosed) of PTSD in Cambodia and among Cambodian refugees in the diaspora.[6] Researchers have also found that PTSD is passed down genetically and socially from Khmer Rouge survivors to their children.[7] Um conducted an extensive series of oral histories with Cambodian American survivors of the Khmer Rouge, "1.5" (first-generation immigrants who moved to the new country in childhood) and second-generation Cambodian Americans, and Cambodians who remained in Cambodia. Trauma is a shared, collective, national, and transnational experience and also a deeply differentiated and personal one. Her book gives rich historical background on the origins of suffering and catalogs the experience of living through genocide and rebuilding lives when "the sequelae of trauma that remains is [sic] neither fully present or fully absent."[8]

There is also substantial criticism about the applicability of the Western concept of trauma to Cambodian survivors of the Khmer Rouge, both because of its mismatch with the cultural context and its pathologization or essentialization of people who have experienced painful events. One response is Chhim's development of a more culturally specific concept for suffering after the Khmer Rouge. *Baksbat*, literally translated to *broken courage*, is used in Cambodia to express psychological experiences following the life-threatening and terrifying experiences of the Khmer Rouge period.[9] Chhim and the staff of the Transcultural Psychosocial Organization (TPO) have argued that *baksbat* is a culturally specific set of symptoms of psychological distress, which include symptoms of the Western diagnosis of PTSD but also include symptoms that are specific to the crisis of the Khmer Rouge and the cultural environment of Cambodia. For example, people suffering from *baksbat* have a particular fear of authority, due in part to Cambodian cultural expectations for social hierarchy and the particular brutality and control of the Khmer Rouge authority figures.

Though it is important to acknowledge the shaping effects of pain and violent histories, scholars have also pushed back against essentializing people for their trauma. We can learn most about the emotional legacy of the Khmer Rouge from survivors and their families. Chhun writes about her experience living as the daughter of two Khmer Rouge survivors, witnessing the way their memories manifest as physical pain, and being open to the silences of their pasts. She follows a call from Eve Tuck, and, rather than focusing on "damage-centered" research, she narrates how her family, particularly her parents, continue to "walk with the ghost," by being present in their lives and "bearing witness to joy as well as loss."[10] Chhun's narrative works as an "anticolonial feminist intervention" by attuning to the bodies and silences of her parents and to the full dimensionality of her own life and the lives of her family members.[11] I work from her example, in part, in shifting the focus of this book to memory practices that serve to light up joyful affect as well as loss.

Thompson has poignantly described some of the complex miscommunications and power relations between national and international actors around questions of commemoration in Cambodia. She shows the ways that international actors (from NGO donors to tourists) often overlook (or are not trained to see) long-standing Cambodian commemoration practices, which can reside outside Phnom Penh's cosmopolitan settings and reside instead in villages, are connected to Cambodian spirituality, and sometimes are intentionally left to deteriorate according to Theravada Buddhist tradition.[12]

One of these international misunderstandings may be the role of international justice in moving past the Khmer Rouge. Since 2008, the Extraordinary Chambers in the Courts of Cambodia (ECCC) has worked to promote healing through international justice standards of legal redress, by holding the leaders of the Khmer Rouge accountable for war crimes. Within Cambodia, many criticize the ECCC for being expensive, slow, divisive, and starting too late.[13] Hinton's phenomenological investigation of the ECCC suggests that the transitional justice imaginary interacts within both local realities and international power structures, landing in complex ways within disparate audiences. Hinton concludes that, while offering some "limited set of benefits and possibilities," "transitional justice may not necessarily penetrate far below the surface."[14] Schlund-Vials argues that the ECCC exists within an international response that continually fails to hold accountable US foreign policy for its role in the genocide.[15]

Apart from these internationally legible responses, Cambodians have been working, collectively and individually, since 1979 to heal psychological distress through many physical, psychological, and religious approaches. These include coining, cupping, Tiger Balm, massage, moxibustion, religious practice, visiting a traditional healer, fortune telling, and talking to friends, family, and community members.[16] These approaches predate but are resonant with the growing scientific exploration of the embodied nature of trauma.[17] Eisenbrook shows how these traditional approaches often attempt to restore balance in social relations and with spirit relations following the conflict.[18]

Anne Guillou, an anthropologist who studied Cambodian systems of resilience in villages in the 1990s, writes about how she observed, after the trauma of the Khmer Rouge, Cambodian people gradually becoming more involved in rebuilding their lives through such a restoration of spirit relations.[19] She suggests that the "switch off/switch on" mechanisms of alternately forgetting and remembering are key to understanding post-conflict trauma and memory in rural Cambodia. *Neak ta* (literally translated as "person-grandfather" but can more broadly be understood as "village spirits") is essential to understanding this "switch on/switch off," dualistic approach to mourning and loss.

Ang Choulean explains how *neak ta* (which can be seen throughout Cambodia) is a distinctive part of Cambodian religious practice and a ménage of transnational influences from India, China, Indonesia, and various forms of Buddhism, Hinduism, and animism.[20] A *neak ta* is a village guardian spirit and, as Choulean describes, is "two in one." One of the two things is the "soil of the village community," which means houses, rice fields, and other spaces used by the villagers. The second thing is a human, but it need not be a specific person and sometimes represents a legendary figure. *Neak ta* can take different forms like a tree or a stone or even termite mound, but all of them represent the unification of soil, which is associated with rice cultivation and people. Massive bereavement has been incorporated into the popular religious framework of Cambodia, particularly into the cult of the *neak ta* with whom the dead of the mass graves from the Khmer Rouge share many characteristics. This kind of bereavement allows a specific relationship between the dead and the living and a memory of the genocide that consists of alternately forgetting and remembering.

Acts of commemoration in Cambodia are not only integrated into the built environment but also developed through practices of art and craft. Uk describes the ways that the Jorai, an indigenous group who primarily live in remote jungles of Northeast Cambodia, have developed practices of resilience that have allowed this group to survive despite the ruptures of American bombing during the Vietnam War, the Khmer Rouge genocide, and the long-standing civil war that followed these events.[21] She particularly focuses on the role of craft as a practice of resilience and argues that objects and the act of crafting provide a new lens for understanding how post-conflict communities interact with their past.

These responses through art and craft are not confined to Cambodia's national borders but are now created in the diaspora as well, reflective of the transnational origins and impacts of the history of violence. Cambodian American authors have analyzed the ways that diasporic Cambodians have responded to the genocide.[22] Schlund-Vials has written extensively about the art production of the Cambodian American 1.5 generation for culturally specific genocide remembrance, by artists such as Socheata Poeuv, Loung Ung, Chanrithy Him, Prachh Ly, and Anita Young Ali.[23] Ly situates *baksbat* in the visual cultures of Cambodia and its diaspora, analyzing artistic remembrance of Khmer Rouge histories through case studies of the artists Amy Lee Sanford, Both Sonrin, Rithy Panh, and Sarith Peou. He also interprets the work of artists who address the American bombing of Cambodia, including Vandy Rattana, Leang Seckon, and Chanthou Oeur.[24]

Building on this literature, my concept of infrastructural restitution explores how practices of memory in Cambodia occur on and through media, the communication outlets or tools used to store and deliver information and data. Our memories are highly connected to the media (such as photographs and film and digital media) that we use to record, archive, and disseminate them. Freud was among the first to consider the relationship between psychic memory and material reproduction of memory in media, and suggested that media technologies shift our understanding of memory and storage away from the human brain.[25] Walter Benjamin deepened the link between media and memory, as he discussed the implications of early (1930s) mechanical reproduction of art. Benjamin argued that viewing the "mechanical" reproduction of art fundamentally changes the viewer's experience of that art.[26]

Later thinkers, including Marshall McLuhan and Frederic Kittler, expanded and elaborated this idea of memory-media determination.[27] McLuhan argued that media are "extensions of our human senses" and that there is a "psychic message from the medium." Kittler similarly argued for the determination of media and delineated the tight linkages and representational relationships between memories and media. He argued that our bodies are extensions of technologies and that technologies represent our bodies. Kittler also tied media to the supernatural and argues that media allow "memories and dreams, the dead and ghosts" not only to live but to become technically reproducible. With the new technologies, Kittler argues that the "realm of the dead is as extensive as the storage and transmission capabilities of a given culture . . . In our mediascape, immortals have come to exist again."[28]

Painful memories, like ghosts, are uncertain, characterized by the registers of the emotional, magical, and subjective; likewise, scholarship on noise has highlighted the ways that media are fundamentally uncertain. This body of work emphasizes the ways that media degenerate, change with time, and become filled with error and glitches. This noise plays an important role in the messages we receive about the past. As Parikka has argued, "noise, not meaning, is often the focus of our technical media."[29] There is never certainty that the message we send is the message received due to the possibility of unplanned noise, the result of the technical properties of the machine we use to send the message.[30] Noise represents the "uncanny and alive" qualities of the technical medium outside the bounds of systematic control.[31]

Noise can act as a point of creative inspiration. Kelly describes the experimental music genre *glitch*, in which an artist breaks or manipulates media technologies within performance.[32] He defines *cracked media* as the media playback tools stretched beyond their intended purposes and a *crack* as a "point of rupture" ripe for new creative possibilities. Kelly describe cracks in a variety of forms such as the gentle manipulation of a vinyl record to the destruction of a CD player. Krapp suggests that glitches give artists a way to make visible and embrace fundamental limitations.[33] Menkman explains that the ways glitches develop meaning are constantly mutating, as media technologies change and as writers, users, or readers interpret them through shifting social, aesthetic, and economic dynamics.[34] Poor-quality ripped films are common (perhaps particularly in low-income countries) and have value for many reasons, including for displaying "defiance and appropriation."[35]

I argue that noise-ridden media associated, but separate from, painful memory can help with the processing of painful national history. The history of Cambodia's conflicts gives the errors, glitches, and poor images in the films exhibited by Preah Sorya a particular meaning and importance. The films are full of *disintegration noise*—they are partial, unedited, or show grainy marks of past formats, due to their condition of being "lost" and then recovered.[36] This case shows the limits of original materiality and a moment of creation and, in so doing, the nature of the relationship between media and memory as it moves through time. Media in this case do not simply and constantly constitute memory, but do so partially through uncontrolled and unintended mechanisms of noise, which change over time. The meanings granted to noise are also constantly changing within shifting political and social dynamics. Disintegration noise, like other kinds of error, is unintentional, though it can be embraced as a catalyst for new forms of creativity.

The noise also provides an opportunity for healing. Preah Sorya actively commemorates these films in order to move away from the painful past into a better future. Though the films do not directly refer to memories of war or genocide (in ways that some art critics or tourists look for in commemorative contemporary art in Cambodia), this project celebrates artists who died. Preah Sorya commemorates them by focusing on what they believe are the most beautiful parts of Cambodian history: artist role models and their way of life that has subsequently disappeared. The films are not remembered sitting in a hard drive, but they are actively remembered, rewatched, and played during in-person events. This case gives insight into how historical media are artifacts of active commemoration and sites for dreaming forward. Their disintegration noise thus helps us understand the complex relationship between media and memory, and how positive-affect laden media still associated with violent pasts allow us to process painful memories. Rather than representing past events or states of consciousness, media are rather an emotional access point to histories of violence and provide space for healing.[37]

A few weeks after the film festival, I went to the house of Preah Sorya, located on a small alley a kilometer from the Royal University of Phnom Penh. Virak greeted me at the metal gate and asked me to take my shoes off and park my moto in the main room of the first floor of a vertical townhouse. There were sixteen people in the group, and many of them lived together in this house, "like a family," one suggested. Since many of them were students and came from the provinces to study, this home was a good place for them to live in Phnom Penh.

I followed Virak and climbed six flights of stairs to the roof. I arrived around 6:30 pm, just after sunset, and a glimmer of light remained around the wide horizon. Coming from (increasingly) congested Phnom Penh with skyscrapers filling the cityscape, the view from this roof in this neighborhood (near the airport) looked vast and uncluttered. They had a huge balcony on the roof overlooking a green, swampy field. The group recently built twenty benches and set up a projector in the back ("high definition—we bought it in the provinces," they told me proudly). An eight-by-five-foot screen was set up in the front of the room. I asked about the history of this makeshift cinema and they told me that they built it within the past few months and that showing this film to me was their first "public" screening. They had had showings before but they limited them to the "Preah Sorya family" and other family members.

The group posted about this screening immediately on Facebook. They shared a lot of old film material on Facebook, as well as information about their screening events. I asked them about the benefits of the Facebook platform for their group. They said that it was easier to share movies with their peers on it. Vandy said his friends are "the new generation in Cambodia" and wanted access to clips of these films in high quality and that he wanted to create easier access to them. Normally, they posted clips and film montages a few times a week, though full films were not generally available on their page.

That night we watched a series of three films that came an original film reel from a collector's shop in Kampong Cham province. The three films were all partial versions and were stuck on the same reel they found in the shop. The members of the group had watched all of these films a number of times already. Vandy later told me, "I watch these original films many times because I want to remember (*chong cham*). I watch them again and again because they make me feel (*mean aaram*). When they cry, I cry . . . when they are happy, I am happy (*kay sabbaay, khnom sabbaay dei*). I watch a movie, for example, four times and it still looks really good. The film plays (*lang*) the daily life (*jiwut*) of the past. Watching it is exciting. . . . For a [contemporary] ghost movie, one time is enough because I get scared. I like to watch these movies again and again. The stories, the special effects . . . they make me feel."

We started with the first movie made by Ly Bun Yim, one of the most prominent producers and directors from the 1960s Cambodian film scene. As the film began, Virak explained some background about the film to me: "It was one of the first color films and one of the first films connecting sound to image in Cambodia." Yim also starred in it as an actor; his father-in-law played the humorous villain; and his first wife played the love interest. Yim's first wife, Virak explained, died in the Khmer Rouge era, and their only child now lived in the United States. One of the shots took place at Wat Ounalom (still a standing and popular wat in downtown Phnom Penh)—and some of the students watching in the back shouted, "Ounalom!," visibly happy to see the familiar landmark in an earlier age.

The film cut out about halfway through and another film began. This one was called *Golden Violence*, starring Kong Sam Oeun and Kim Nova.[38] As Nova appeared on screen, multiple students came up to me and described her as "one of the beautiful Khmer women." A girl in the back murmured, "Ooooh, very handsome!" (*saart na*) with the first close-up of Sam Oeun. The film was made in 1969 and no remaining full version exists; this film

Figure 5.6
Preah Sorya cinema and screening

was an unedited early version. Some shots repeated over and over with slight differences based on different takes. The film's effect was eerie and mesmerizing, an abstract collection of clips showing beautiful places and a different time in Cambodia. The color of the film reel was vibrant. One series of shots that lasted for over twenty minutes showed Sam Oeun and Kim Nova sitting on a paddleboat riding around Boeng Keng Lake wearing "modern" outfits.[39]

The third film of the evening was a comedy from 1968 (made by the same director of *King Kong*, Virak pointed out to me). The film was a slapstick comedy about marriage. It differed from other films I've seen from this era in that it didn't have a modern feel; the movie was set in rural Cambodia with most scenes taking place in a typical wooden, two-story Khmer house. Virak explained that this director made films in a studio in Kandal province (a few kilometers outside Phnom Penh) with four hectares of land, giving them an authentic rural feel. By the time the film ended and I got ready to leave, eight students had gathered to watch the end of the film, as well as an older woman in her late twenties and her toddler daughter.

I walked downstairs and as I started to get ready to ride out, I noticed that the group had converted the entryway of their home into a makeshift archive. In this main room, I saw an old projector, cassettes, and lots of posters and ephemera from 1960s films. Virak explained in an earlier interview that the group liked having control over the films. "If we want to watch a film, we can do it," he said, and they wanted freedom for their organization. They had an extensive collection of magazines and novels from the 1960s—printed and copied inside plastic folders—that they found at collectors' shops. They also had a stack of old film boards and advertisements. Virak showed me an ad from the first international film festival in 1964, which he found in an old magazine. He had information about films in historical magazines in French, Chinese, English, and Khmer. He told me that one of their goals for the future was to make a more official physical archive.

I was grateful for being invited to Preah Sorya's oasis of film history, which existed hidden within the rapidly developing city. They shared with me their broad and deep knowledge of the history of Cambodian film, its actors, producers, and technologies, which they had been able to glean through their artifacts and research. They knew and celebrated the actors who had died, left Cambodia, or stopped making art. Through the films, even though they are spotty and partial, the group is able to see and show images of prewar Cambodia, including both its modernist urban aesthetics and traditional ways of life. They could share bits of this imagery through Facebook, on which they give frequent glimpses of the work they do and the resources they have.

The case of Preah Sorya highlights the ways that media technologies change over time due to deterioration and the layering of history, emphasizing the unpredictability of how media constitutes changing and unreliable memory through time. The films are full of *disintegration noise* such as spots and blurriness from material deterioration, the visible layering of formats on top of each other, unedited scenes, soundlessness, and partiality. In *Neeung Jew Kreu Fa*, for example, we see three formats of film all collaged on top of each other (film, VHS, digital), and the past becomes radically visible in the present through materiality. The flecks of the original film reel, the shaking of a hand holding a camcorder, the black and white streaks of VCR screening, and the moments of digital watermark—all of these are visual reminders of the layers of history in these films and the way memories are filtered as they are transferred. The members of the group watch these films over and over again.

Disintegration noise contributes two ideas to theory of media and memory. First, it suggests that noise can be an asset in old media in relation to memory: rather than highlighting media's representation of a past reality, it demonstrates media's status as old and, therefore, valuable. Second, it gives the opportunity for new creative possibilities and allows the memory media to help Preah Sorya experience positive affect around the history of the country and build promising a new future. I therefore return to the two understandings of how media act as mechanisms for accessing and storing memory with which I opened the chapter. I show that these films both document and reconstruct the past, while also acknowledging its uncertainty (made visible through noise). The noise and uncertainty give room for the hope of healing.

Material representations of memories in this context are scarce and deteriorating—just as these films are. The disintegration noise therefore does not detract from their value but instead highlights their rarity. The films are important because they are rare and envision a former time in

Phnom Penh. Though many Cambodian young people have told me that they know some things about the country's history through school classes or talking to relatives about their experiences, many also tell me that it can be very difficult to talk to their parents and other relatives about the depth of their experiences and that they have a shallow understanding of the wartime and before. Preah Sorya collects as much as they can to learn about Cambodian film history, including 1960s films, journals, songs, and other ephemera, and then they repair, digitize, and redisplay them in media formats that are digestible to their peers in film screenings and on Facebook. Preah Sorya loves these films because they enhance their knowledge of the prewar time; they can see actors who died during the Khmer Rouge era, Phnom Penh before it was evacuated, and a way of life that ended during the war.

Excavating and reimagining the Phnom Penh shown in the films is a process that will never be complete. Though the group has tried to gather as much information about this era as they can, there is still a deep sense of the unknowability of the pre–Khmer Rouge era. The limits of what the group sees on screen correlate to the limits of the memory of that time. The group members can't know what the off-screen world in Phnom Penh was like when these films were made or even how most of these films end. Due to the limited number of films available and the value of each, Preah Sorya constantly repeats them, both in public film screenings and by sharing and resharing parts of them on Facebook. Vandy said he could watch the same movie four times and still *feel*.

The disintegration noise in the films also gives Preah Sorya a creative opportunity to fill in gaps by putting together the media that they have with other creative forms. The group combines the partial forms of media with in-person presentations of aging actors and reenactments, leading to an exciting new creative output. They literally finish the story of unfinished films through their own lip-synced rendition of a period song dressed in period clothes. Like self-described glitch artists, Preah Sorya, through all of these activities, embraces rather than ignores the imperfections of their historical media resources in order to realize new creative possibilities. They make the best of what they have and put together a fantastic viewing experience through inventiveness and imagination from (what could be understood to be) poor quality and partial films.

This case therefore points to one of the core arguments of this book: that infrastructural restitution is an avenue for accessing positive affect about the past amid an overwhelmingly painful history, and thus acts as a vector for healing. The group focuses on memories that allude to the trauma of the wars without directly referencing them. The group focuses not on memories from the traumatic event (the Khmer Rouge and wartime) but instead on the positive cultural outputs from the period before that. The students are able to commemorate lost artists without focusing on violence; they thus decenter what they often perceive to be a simplification of their national history through an international focus on the Khmer Rouge period.

Cambodian young people often tell me that they feel trapped and made one-dimensional by the force and darkness of their national history. They hope to generate optimism and a feeling of creative empowerment among their peers through the dissemination of knowledge about the cultural "glories" of the past.[40] In so doing, however, this case also showcases another theme of this book: these commemoration practices are not neutral. The period Preah Sorya romanticizes was one of deep inequality. What they choose to focus on (the 1960s and early 1970s film culture) was also deeply embedded in imperial and elite politics. Some readers of this deeply nostalgic project could also see a problematic subtext of ethnonationalism.

The creative ways that the students collect, repair, and display these films is done for *their* generation as a form of dreaming forward. The group wants to know about the past for themselves and for their friends—it excites them to see these historical cultural outputs. We can see the targeting of a young, urban, and educated audience in the choice to disseminate and discuss these memories through their active Facebook presence. The films provide an alternative Phnom Penh: the past gives them artistic role models and images of a Phnom Penh full of natural space and youth opportunity. They want to build a Cambodia that encourages more artistic production and integrates some of the best parts of this pre–Khmer Rouge society.[41]

These films also allow them to dream forward toward an emotionally healed Cambodia. Other scholars cite the important links between memory, violent histories, and hope. As Crownshaw and colleagues argue, memories of violence must also be accompanied by the "remembrance of the future" as a way to move toward hope.[42] Memory is a site for the reevaluation of identities in a postcolonial and post-conflict world.[43] Harvey comments that the "power of collective memory is politically very important, provided it is connected also to the notion of desiring something different."[44] In an arguably more modest intervention, Chhun's anticolonial feminist approach is to acknowledge how people can live full lives while "walking with the ghost." This project allows the students to move past trauma, past *baksbat*, and past violence. In serving to alleviate the "problems of the way of the heart," watching these films gives the group an emotionally smoother way forward.

6
MAKING MEMORY LIVE: INTERNET TOOLS OF CAMBODIAN MEDIA HISTORY

> The archive was not a building, nor even a collection of texts, but the collectively imagined junction of all that was known or knowable, a fantastic representation of an epistemological master pattern, a virtual focal point for the heterogeneous local knowledge of metropolis and empire.
>
> —Thomas Richards, *The Imperial Archive*

Amazing Cambodia is a Facebook page that disseminates found, curated, and sometimes reconstructed photos from the pre–Khmer Rouge era.[1] This digital tool responds to a particular need in Cambodia for finding, restoring, and disseminating audiovisual historical material in Cambodia, since much archival material was lost during the Khmer Rouge period and ensuing wars, and institutional archives in Cambodia still encounter many challenges, including lack of government support, external funding, and familiarity with the technical requirements. This lack of institutional archives has led many Cambodians to seek alternate resources for learning about and getting access to historical media forms. Sokmean Srin is the independent researcher who runs the page. His page is an important source of information for many student groups as well as professional Cambodian and foreign researchers. The page has over 96,000 likes on Facebook.[2] When Srin posts historical remnants on his Facebook page, they can get hundreds of likes and dozens of "shares." He creatively constructs the page from historical research, research travel around the country, original photography, and contextualization.

This chapter explores ways that new media creators in Cambodia are using online platforms as repositories for historical media artifacts. I show the ways that these media creators construct these repositories through

Figure 6.1
Srin at pagoda near Oudong, August 2018

work that bridges online and offline spheres. Much of the offline work is place-specific and invisible to the globalized audience of the internet platforms on which they publish. The rendering of this invisible offline work emerges from a legacy of the imperial archive, the colonial fantasy of consolidating authoritative knowledge, while systematically excluding some forms of place-specific knowledge. The use of contemporary online platforms for these historical projects solidifies ties between people and places (as of yet) lightly connected to centers of technology with emerging tools of globalized techno-capitalism.

This chapter analyzes and contrasts the creation, use, and promotion of two internet-based platforms that store and disseminate historical Cambodian media. The first is Amazing Cambodia. The second, App-learning on

Khmer Rouge History, is a mobile application run by the Bophana Center, a nonprofit audiovisual archive and training center in Phnom Penh; the European Union and the Japanese REI Foundation funded the application initiative. The data-rich application teaches lessons in English and Khmer about the Khmer Rouge period (1975–1979) and also stores videos, songs, and images from this period to support historical learning.

These two tools are in ways oppositional to each other; the first is grassroots and built through volunteer labor, whereas the second was developed by an international NGO and funded by foreign grant-giving bodies. They also take starkly different approaches to commemorating those lost in the Khmer Rouge. Yet I argue that they similarly show the politics and material conditions for creating a digital media "archive" in Cambodia, and what kinds of knowledge are commonly left out of these new globally accessible repositories of historical material. I argue that these two tools construct collective memory by actively and deliberately collecting, posting, and reposting historical media. Sometimes this work happens online, often through the mechanics of online sharing. Through researching, repairing, posting, and disseminating historical artifacts, the creators of these two tools enact what Chun terms the "enduring ephemeral."[3] These cases, however, deepen this concept by demonstrating that making this information matter includes significant dialogue and work within the offline world. I describe the empirical object of the "provincial trip" as an illustrative example of the work of movement that goes into the construction and dissemination of a curated form of memory on these platforms.

In bringing Cambodian provinces closer to the city and other centers of information and technology, I also argue that the developers of these new online tools contribute (perhaps inadvertently) to a new form of the historical project of global consolidation of knowledge.[4] Through the provincial trips I describe, both Srin and the Bophana Center team bring tools of modern global techno-capitalism (smartphones, Facebook, internet, projectors) into places still marginal to entities who control information (the state, multinational corporations, and international NGOs). In this way, they bring international knowledge to the margins and make (some kinds and textures of) marginal knowledge globally accessible. The specific platforms on which these tools run—Facebook and the smartphone mobile application—play important roles in *how* these tools act as archives and connect regional knowledge to technological centers of power.

The empirical descriptions in this chapter emerge from multiple interviews with the creators of the new archiving tools in Phnom Penh; discussions with users; my own use of the tools; provincial trips for collection of materials to Kampong Chnnang, Kampong Cham, and Kandal provinces; dissemination and promotion for each tool; and participant observation in related events (such as the application kickoff) in the summer of 2016 and from June 2017 to January 2019.

Amazing Cambodia is a Facebook page and eclectic archive of Cambodian media remnants with an emphasis on the 1960s Sangkum Reastr Niyum period, particularly the Khmer traditional-modernist aesthetic characteristic of that period in Phnom Penh, including digital images or clips of film, photography, architecture, music, and radio of that period. The content includes, for example, old photographs of 1960s or early 1970s New Khmer Architecture, comparisons between "then" and "now" of specific locations from before and after the Khmer Rouge period, and photos of famous historical events and people. Srin also comments on important cultural figures from the Sangkum Reastr Niyum period who survived the Khmer Rouge regime; for example, he wishes them well during birthdays and posts death announcements. His text is in a mixture of English and Khmer.[5] Many of the photos are watermarked with www.facebook.com/AmazingCambo or with the photographer's name.

Srin is originally from Phnom Penh and graduated in 2014 with dual bachelor's degrees in tourism and English from the Institute of Foreign Languages at the Royal University of Phnom Penh, and was teaching English as a second language during the course of my research. In 2018, he told me that he enjoyed doing the research for the Amazing Cambodia page—so he did it in his free time. Up until that point, Srin had run Amazing Cambodia for five years. He told me that he gets documents from institutional archives (including the National Archives) and private collectors. However, some of the government-run archives are not easily accessible or convenient to sort through.

Cultural celebration was important to what Srin wanted to build with Amazing Cambodia. He told me that he strove to find and post "anything interesting, anything rare" in Cambodia. Referring to the recent building construction boom in Phnom Penh and across Cambodia, he lamented that "when a country becomes rich, it seems to want new buildings." He wanted to understand and preserve the old. His goal and his passion is

MAKING MEMORY LIVE 163

commemoration. "We the young generation should do research on people who passed. It is our responsibility." The site gives him an unofficial but powerful mode to communicate the results of his research.

In the summer of 2018, Srin was regularly traveling to provinces to visit old pagodas to "fill in the missing information" about them. For example, he would ask a monk or other villager who lived around the wat about the social history of the pagoda. He asked about what happened at these sites during the Pol Pot regime. He learned that many of them, for example, became jails or hospitals. Srin told me it was often hard to find written documents to contextualize his research.[6] He thoroughly combed through all the available materials he could find for old photographs. He went through the official, analog archives and all the websites collecting and disseminating old images of Cambodia. Often Srin found photos without accompanying context and had to discover what they meant through clever research.

In August 2018, Srin and I went on one of his typical research trips together, this time to Oudong, a former royal capital (from 1601 to 1866) in Kandal province, about an hour from Phnom Penh downtown by tuk tuk. The goal of the trip was to capture photographs of a pagoda nearby the main temple complex and to post them on the Amazing Cambodia page. After arriving in Oudong, we stopped for breakfast. Sitting over fried rice and noodles, we heard a Khmer oldies song called "Stung Sen Paris" on the loudspeaker. Srin told me about it. A Cambodian refugee who moved to Paris in the early 1980s wrote it. She was looking at the Seine, thinking about the Stung Sen river in her hometown of Stung Treng, and feeling sad and homesick.

Srin continued, "When I was in second grade, I would regularly see a bicycle come around my neighborhood. The bicyclist sold pulled candy and played oldie songs (like this one) on a loudspeaker attached to the bike." I told him I had never seen a candy seller like this, and Srin told me that, in the countryside, these carts are still popular, though in the city they are no longer common. Inspired by the bike, he started collecting old song lyrics. He told me, "I started to compile a book of song lyrics in grade five. On the side of the road, there were small shops that have the lyrics printed with images. These images are sometimes pictures of the singers. They are also sometimes symbols of the themes of the songs. If it is a sad song, then it could be an image of a sad woman with tears. Sometimes they are images of singers when they were alive [if they died during the Khmer

Rouge period]; they often resemble drawings in comic books. This was a big trend in the 1990s. I bought many and would like to color the images in."

He then told me about his upbringing. "My maternal grandmother has always been my guardian.... I live with my two younger brothers.... One of them is getting a BA certificate in tourism and one is still in high school. We also live with my mother's sister. My grandmother is now about 70 years old." I asked him if his grandmother often told him interesting stories about the time before the war. He said,

> She studied in a high school near Bong Keng Kong in Phnom Penh. She tells me a little bit about Phnom Penh but not much about the rest of Cambodia. She tells me often that we, the young generation, are very lucky because we can take trips to the provinces. During the wartime, it was very difficult to travel because of the road quality. She was married young and dropped out of high school. Her husband bought her a Peugeot 404—she always tells me about this car. She loved it. Her husband worked in the railway station and would go to the Thai border for work on the train. When he traveled up there, he could buy her new clothes and makeup, which traveled over the border from Thailand. My grandmother's fond stories of the car make me excited to travel to different provinces now to study history.

Srin's grandparents and mother were transported to Battambang during the Khmer Rouge period. His grandfather passed away there. Srin's mother was born in 1967 and was a child during the war. His father's family is from Kandal province, but Srin doesn't know them. He told me he wants to understand more from his grandmother but thinks it can be a sign of disrespect to ask older people questions about the past.

We visited some pagodas near Oudong then went to the market for lunch. Sunday at noon in Oudong market was busy and frenetic; there was a large variety of market sellers and middle-class visitors coming from Phnom Penh. It was rainy season and the humidity was extremely high. Srin ordered fresh fish balls. The vendor took spoonfuls of green paste and threw them into a vat of hot oil. We walked to a wooden plank where we sat for lunch. Srin offered me one of the fish balls.

While in the middle of eating, a pulled-candy bicycle vendor like the one Srin had described to me at breakfast rode by our picnic spot. The music of the oldies song blasted all around us. I got up in the middle of the meal to run after it. The rider stopped a few meters away to sell candy to a small kid. After waving and saying hello, I went back to our picnic spot and told Srin how excited I was that we saw the same pulled-candy vendor that

Figure 6.2
(Left) Seventeenth-century braangk; (right) pagoda from 1964

he had described. Srin laughed and nodded agreement. "Do you want some candy?" I asked. He said, "I don't like the candy anymore. It is so tough, hard to pull and too sweet. It hurts my teeth now."

As soon as we left the Oudong market on our tuk tuk, it began to rain. Through this heavy monsoon rain, we took our tuk tuk to visit the site of the original royal palace during the Oudong period, which has been reconstructed into a contemporary Buddhist temple. I felt a part of the emerging middle-class urban tourist experience, exploring Cambodia's heritage on a Sunday, and reflected on the importance of royalty and the religious order in Cambodian society. I appreciated the green space and being outside the urban environment of construction, congestion, and pollution in Phnom Penh. Inside the temple, there were twenty large buddhas and a man sitting to pray. The only remaining artifacts from the Oudong period were original cannons surrounding the temple from the seventeenth century.

Srin took photos on his smartphone of the various sites we visited. He told me, "I sometimes copy the files into a computer when I can but the phone is more convenient. Mostly I store all of my images on my phone. I had a phone stolen just after the previous Pchum Ben [ancestor worship day, September 2017] and I lost many images. I went last year to this very site and took photos but lost the images when my phone was stolen, so I could never post them on Amazing Cambodia."

We finished our research day at a final historical site where we saw three buildings: a braangk (temple from the seventeenth-century Oudong period,

Figure 6.3
Colonial-era (1920s) storeroom

which famously held Buddha bones in the colonial period), a pagoda from 1964, and a colonial-era building (likely from the 1920s), which is now being used as a storeroom. Srin showed me on his cracked phone a colonial postcard that he found in a book collection and that depicts an earlier image of the buildings on this site. The colonial postcard was not labeled with the right site but he has been able to label it correctly from the distinctive image of the braangk. He then showed me another photo of this pagoda being inaugurated in 1964, pointing out the image of the braangk. He found this second photo in the Buddhist magazine *Kambuja Sorya* at the Buddhist Institute in Phnom Penh.[7]

We walked from the pagoda site down a small dirt pathway next to a leafy green pond. When we arrived at the end of the pond, we are able to see the sites at the exact angle of the two old photos that Srin had found. He held up his phone showing the 1964 inauguration photo from *Kambuja Sorya*. Standing in front of the pagoda, across from the pond, he was at exactly the same viewing angle as in the historical photo. He beamed.

Figure 6.4
Colonial postcard on Srin's phone

Figure 6.5
Image from *Kambuja Sorya* of 1964 Oudong pagoda opening

Figure 6.6
Srin at pagoda, August 2018

Srin took his own photo of the scene and told me he was very excited to make the then-and-now-shot of this site. He was disappointed because two trees have grown in front of the pagoda and it is difficult to make out the pagoda building as clearly as you can see in the 1964 photo.

Chun investigated how our shift from analog to internet-based archiving changes the relationship between storage and memory, both the mechanical memory of computers and the human experience of remembering.[8] At the onset of internet archiving, enthusiasts believed in a promise of unlimited storage. As more scholars and practitioners illuminated the true environmental costs of data (including the huge energy cost of data centers and so-called cloud computing), we have slowly become disillusioned with the idea that data can continue to grow infinitely and that all historical information can be digitized and made universally accessible.[9] Apart from these material limitations, we have also realized how easy it is to lose information on the internet. As Chun describes, keeping a photo in a server does not mean it is remembered. Chun instead shows how the long-standing conflation of storage and memory has obscured how digital memories must be stored on the internet *actively*.[10] The way that we archive on the internet (particularly on social media) involves either constant refreshes or constant degeneration. This is due, in part, to the pace of social media, which are based on algorithmic ordering for ideals of freshness and relevance.[11] These practices bring to the forefront both the permanent *and* the ephemeral qualities of digital archiving. Memories are permanent only if they are continually shared, and this is the heart of what Chun calls the *enduring ephemeral*.

The kinds of knowledge shared and the ways this knowledge is constructed and disseminated in Cambodia deepen and expand Chun's concept of the "enduring ephemeral." Amazing Cambodia does not maintain itself as a source of global knowledge and memory; instead, it is the product of individuals, who work both online and offline, actively putting together and refreshing these tools. In this case, Srin is responsible for keeping the ephemeral enduring. The ephemeral endures through frequent sharing of media memories, both online and through in-person provincial trips, an important mechanism for historical data collection and dissemination.

Srin's family background and his personal mementos motivate him to do the work he does. His grandmother told him how lucky he is to travel and do research because the movement is possible—something that was possible for her only before the start of the civil war. Details of the landscape that are specific to Cambodia—such as the pulled-candy bicycle—remind him of the country's culture and past and his own childhood remembrances of discovering this past. Doing research for Srin means savoring the particularities of the places he goes. This always means eating special foods in the provinces to which he travels and otherwise finding and participating in what makes any region unique and special. Because of his thorough knowledge of available historical resources and the countryside of Cambodia, he is able to puzzle out connections between the past and the present. He is able to recognize, for example, the braangk of Oudong from a mislabeled colonial postcard. The final products disentangle the lived experience of knowledge collection and dissemination from the historical visual documents that are preserved. Many of the sounds, smells, and tastes of the place of knowledge are removed by the forms and conventions of the platforms on which these materials appear. The format of the Facebook page means that his followers are limited to the sight of the two-dimensional photographs and brief descriptions that Srin attaches to them.

The Bophana Center is a Cambodian media archive and youth audiovisual training center. Rithy Panh, a Cambodian French filmmaker (who made *The Missing Picture* and *The Land of Wandering Souls*, discussed earlier), founded the archive in 2006. After his family died during the Khmer Rouge regime, Panh left Cambodia for a refugee camp in Thailand at the age of sixteen before moving to Paris. For Panh, in order for Cambodia to move forward, it must confront visual memory of its painful past. The purpose of the center is to provide a collection of material that preserves the cultural heritage of Cambodia and to connect Khmer youth both to this depth of creative history as well as to its history of trauma. Through this exposure, as well as technical skills training, the center's leadership hopes to inspire new forms of creation in their students. Bophana actively uses social media to market their activities. They have videos on YouTube, a lively Facebook presence, and have twice-weekly email marketing campaigns.

The director of the Bophana Center in 2017 was Chea Sopheap. Sopheap explained to me in an interview that film is a particularly important historical artifact because it gives us "some direct experience—we see through film how people look like, how they *feel*." He explained how film is very different from history recorded on paper, which describes things "using adjectives." He explained that the oldest footage the Bophana Center has is from 1899, and that in it you can see what people were wearing and how they lived.

In 2017, Bophana Center was shifting from a curated and spatial/architectural experience of watching historical film and listening to music toward an experience of historical media distributed on a mobile application, the App-learning on Khmer Rouge History. This shift was, in part, encouraged by foreign funding bodies who (according to Sopheap) find new media to be "sexier" than Bophana's more traditional archives. The Japanese REI Foundation and the EU have provided funding for the application, which focuses on Khmer Rouge history rather than broader Cambodian cultural heritage (such as the historical artifacts from the 1960s that Srin loves). Bophana hired historical and IT experts as advisors who developed the application that combines a written history of the Khmer Rouge with video footage and photographs. Sopheap said that they are excited about moving into innovative multimedia applications.

Throughout 2017–2018, the Bophana Center website and Facebook page shifted the image of the center from one that focused on storing and restoring analog film reels to one based on distributing information about the Khmer Rouge using a mobile application. During this period, the cover photo on the institution's Facebook page and website was an advertisement for "app-learning." The advertisement was a gray-scale cartoon that depicted a crowd of people walking, presumably away from their homes during the Khmer Rouge relocation in 1975. White block text streaks the image: "App-learning on Khmer Rouge History" is written in Khmer script with English below. On the right side of the image is a gray badge hat states: "App-learning Democratic Kampuchea History." Two black stick figures are in the middle of the badge; one is more anthropomorphic than the other and has a red scarf around its neck. The other is a small squiggle with a round head, representing the child of the larger stick figure. The two figures together spell out K-R.

In July 2017, I attended the Bophana Center reception for the release of the new application. There were speeches from representatives of the EU and REI Foundation. They dedicated the app to the victims of the Khmer Rouge, some of whom were at the event. The Center set up the event with different kinds of mobile devices and tablets for testing out the new application. Participants, particularly younger ones, took the opportunity to see what the new application looked like.

The application runs in English and Khmer. It is organized into eight parts. The first is "How the Khmer Rouge Came to Power." This includes a timeline starting with the Communist Party in Cambodia before 1970 through how they defeated the Lon Nol regime. The second through the eighth parts are "The Khmer Rouge Conquest of Phnom Penh," "How Khmer Rouge Framed Everyday Life," "Khmer Rouge Policies and Ideologies," "Democratic Kampuchea Government," "The Security System," "The Fall of Democratic Kampuchea," and "The Route to Justice." These are written in text (with Khmer- and English-language options) and are interspersed with photographs, sound, and video clips, including interviews and historical images, some of which are developed into short documentaries by the Bophana team. These include a short film on "The Super Great Leap Forward" and "Life in Cooperatives." The application is downloaded from the app store onto Android or iOS and has embedded original sound and video. Though some content within the app can be downloaded so it can be seen without an internet connection, most of the content must run on Wi-Fi or data.

Figure 6.7
Still from App-learning on Khmer Rouge History launch party at Bophana Center

Archives are always deeply political, but in Cambodia, because of the history of colonialism, geopolitical conflicts during the Cold War, and a divisive and prolonged period of war, the construction of archives has particularly high stakes.[12] The urge to create an authoritative archive—one source of knowledge, or the *imperial archive*—has long-standing historical roots and arguably stood at the center of colonialism.[13] Richards argues that, for colonial powers, "[t]he archive was not a building, nor even a collection of texts, but the collectively imagined junction of all that was known or knowable, a fantastic representation of an epistemological master pattern, a virtual focal point for the heterogeneous local knowledge of metropolis and empire." The drive for state control over comprehensive knowledge led colonial administrations to collect more information continually about their colonies; they would map all space outside the bounds of their current maps and confront problems of epistemological confusion emerging from indigenous knowledge forms.[14] These urges led the colonial state to become the central information-gathering entity of colonial life. The colonial state brought all of its subjects into its project as information-gatherers. Richards explains, "So far as the state is concerned, there is no such thing as a non-conducting medium; everyone and everything, consciously or unconsciously, forms part of the state's internal lines of communication."

Of course, the imperial archive was a myth and could never integrate all kinds of knowledge within the colonies. Stoler studies the technical practices and logic of the Dutch Indonesian archive in order to understand their colonial rule more broadly, and the relationship between colonialism and archiving. She shows, through this description of archival practice, how some knowledge was strategically left out. Stoler pays attention to the *structure* of the archive—qualities such as prose style (persuasion, repetition), classification schemes, and categories of cataloguing—rather than the content. For Stoler, this structure is the key site for understanding the epistemological and political anxieties of colonialism and how Dutch colonial leaders reified some kinds of knowledge and expertise as qualified knowledge and expelled others.[15] Colonial (and postcolonial) subjects have noticed and protested both the control over and the strategic eradication of grassroots and indigenous forms of knowledge in the imperial archive. This struggle continues today, and we can see, for example, youth activists in Southeast Asia creating their own archives, to empower their own kinds of knowledge, on their own terms.[16]

Today the original fantasy of the imperial archive, however, has mutated into postcolonial forms, adapted to the internet era. The platforms of Facebook and the smartphone application resemble the imperial archive in the ways they imagine connecting knowledge from margins to centers of power. Technology use and design are encoded with and perpetuate longstanding power relations, including not only colonialism but also corporate exploitation and authoritarian politics.[17] Western technology designers, sometimes in conjunction with NGOs, still build technologies *for* an imagined development subject "other," sometimes under the guise of early information and communication technologies for development (ICTD) initiatives.[18] These programs have been largely critiqued for not engaging with decolonial politics. Critics also point to their ineffectiveness; for example, these projects disproportionately build smartphone applications even though smartphone applications are not widely popular in many parts of the developing world, including Cambodia.[19] Other scholars show vernacular forms of innovation as a contrast to top-down, development agency–led forms of "innovation."[20] Responding to critiques such as Ames's of the One Laptop per Child program, the academic field of ICTD is gradually becoming more attuned to sociotechnical realities and moving away from strict techno-determinist viewpoints.[21]

There is a tension between the imposition of new digital tools and local creative use of foreign technologies and platforms. Local users across the globe often appropriate digital tools in ways that are unexpected to Western technology designers.[22] Often these creative uses of technology respond to various context-specific conditions, such as poverty, socioeconomic inequalities, illiteracy, language, and already-existing infrastructures of which Western technology designers are not aware.[23] Users can determine only so much, however, and the dividing line between the agency of an existing tool (such as Facebook or mobile phones) and the various ways that the tool is integrated into the environment is sometimes unclear.[24] Facebook has given a platform for disseminating historical artifacts that in ways shifts who has the power to process and control "archival" records.[25]

The power of information control once attributed to the colonial state has become fractured and harder to pin down as the first age of colonialism has waned. The impulses of the imperial archive can in the contemporary world be found in new modes of power, including within technology design and dissemination itself. The rise of the digital and online archives has sparked both excitement and fear around new myths of comprehensive knowledge, held within the internet.[26] As we see the structure of the archive move into fractured online platforms, we see new ways that classification schemes and material form structurally eliminate some forms of knowledge.

These new tools co-opt local knowledge collection and dissemination practices into serving the new myth of global comprehensive knowledge (the internet)—or a new imperial archive. On a content level, both Amazing Cambodia and the Khmer Rouge Learning App have a particular take on Cambodian history and curate the information presented to emphasize a particular story of the past. These tools present this particular version of Cambodian history to a global audience, with implications for international relations and the global perception of the country. They also encourage more interaction between technologies of global capitalism, including both the Facebook platform and smartphone applications, for Cambodian users in Phnom Penh and the Cambodian provinces.

The postcolonial politics of platforms matters to understanding the work that both new tools are doing. One of the challenges for technology in Cambodia has been the use of Khmer on hardware and software first created in an English-language setting. Facebook has poorly translated its user interface into Khmer language.[27] This condition makes it difficult for people who use only Khmer to feel as if they are using Facebook safely, because the settings (including privacy settings) are translated in a way that makes it impossible for most Khmer-only speakers to understand. Given the tense political situation and Facebook arrests for oppositional speech in Cambodia in 2017–2018, these conditions make for a very dangerous set of circumstances for new internet users. Insufficient localization efforts by transnational technology companies put some of the most marginalized users at disproportionate information disclosure risk when using new internet tools.[28]

The poor translation of Khmer-language Facebook represents a broader trend: mobile computing platforms often privilege the use of majority over minority languages for a variety of technological factors (such as autocorrect), sociocultural dynamics (such as audience reach), and linguistic characteristics (such as orthography).[29] Translating the text of platforms into minority languages requires work and presents tech companies with challenges that they have, to date, not put sufficient resources into solving.

Translations of scientific terms into native languages must take full account of the cultural, linguistic, and social dimensions of language. The work of appropriate and understandable translation sometimes requires iterative effort and multiple modalities. Two university teachers in Cambodia, for example, translated English scientific terms into Khmer and taught the concepts effectively through English-Khmer language pairs, repetition of terms, and pictures.[30] Not only does software need to be created with useful translations, but the hardware needs to allow for the use of Cambodian characters. Khmer script was added to Unicode in 1999, but many Cambodians still face challenges in typing in the script.[31] Many entrepreneurs have worked to develop a functioning Khmer keyboard for smartphones.

Transnational technology companies based in the West often take a "view from nowhere" approach to designing for a "universal" user base.[32] Meanwhile, activists and start-ups within the tech sector in Phnom Penh work to make space within these communities for preserving and promoting the Khmer language (e.g., see the Koompi computer by Smallworld, Khmer keyboards, and the Wapatoa blog). Inequality, however, remains and it is much more difficult for people speaking only Khmer to be involved in the tech sector. English-language proficiency often comes with class privilege. This is to say that the use of Facebook and, more broadly, new digital tools can be challenging for those of lower-class status.

Facebook works, as does many technology companies, to communicate to their amateur users about the ways the site can fit their needs. However, they also, less obviously, align the product to the needs of advertisers, professional content producers, researchers, and governments. The media platforms claim to "empower" individuals to communicate, implying that they are "lifting us all up, evenly" through user-generated content. This discursive positioning makes it seem that new media platforms catalyze "open, neutral, egalitarian and progressive support for activity" and user-generated content. In this way Facebook (and other large technology companies) and other new media platforms are distinguished from traditional media platforms and obscure the role of advertising and surveillance on these sites.[33] They also obscure the ways that they keep the license to the content in perpetuity, thereby accruing permissions to huge archives of cultural heritage often without clear understanding from the users who uploaded this material.[34] Across my interviews, very few Facebook users understood the business model of Facebook that they so regularly used.

This becomes particularly relevant when Facebook is used as a depository for rare and valuable historical material.

This business model and the socioeconomic and historical conditions that underlie it have led to enormous global inequalities. I flew from Phnom Penh to San Francisco once in 2018 and was struck with how rich the latter city was, and how that wealth was based, in part, on the money that has been made around the world through collecting data that populations often are unclear about. At the same time, Facebook is growing ever more dominant as a market player and being co-opted by regimes to control their populations; they also do little to protect all of their users. These arguably exploitative conditions have led some scholars to call Facebook, and the techno-capitalism it represents, a mode of data colonialism.[35]

Though the App-learning on Khmer Rouge History is independent and does not accept advertisements, to use the app required having a working smartphone and paying for data to download data-rich applications and stream videos and music. Not all of the students that the Bophana team visited on the rural outreach trip had access to these technologies. Further, the goals of learning history and the medium of the app were set in part by international donors' agendas to increase the use of technology in their development support.

In early 2018, the Bophana Center toured the country to advertise the application to rural students. In March 2018, I traveled with a part of the Bophana team—Chetra, Neary, and Toh—to a small town in Kampong Cham province, just over two hours from Phnom Penh and about an hour from the provincial capital. Chetra was a twenty-four-year-old recent graduate of the Royal University of Phnom Penh (RUPP), where she studied history. She is from Takeo province but was given a scholarship at eighteen to study in Phnom Penh ("For kids who come from poor backgrounds," she tells me). Neary was twenty-eight and also a graduate of RUPP, where he studied IT. He worked at a company for a few years before he started to work on the Khmer Rouge History application at the Bophana Center. Toh was our driver. He grew up in a Thai refugee camp and then in 1990 moved with his family to Phnom Penh, where he went to school until tenth grade and then worked odd jobs. He became a friend of Rithy Panh and when Panh made *The Missing Picture*, he asked Toh to help him make the clay figurines. Panh loved the first two figurines that Toh made and they collaborated to make hundreds more for the film. Toh continued to work at Bophana

Center doing different jobs, including spending a few months on the road with Chetra and Neary.

Chetra, Neary, and Toh's mission was to disseminate information about the Khmer Rouge Learning App to students around the country to increase its usage. The promotion team started in Phnom Penh, marketing to thirteen schools in early February. Then they went to eighteen schools in Takeo. Our trip to Kampong Cham was the third trip that they had done for promotion of the tool. After our trip, they were planning to visit eighteen schools in Kampong Cham, ten schools in Siem Reap, and twenty in Battambang.

We left early in the morning from Phnom Penh to go to Kampong Cham, stopping for breakfast on the way. We arrived in Kampong Cham around 11 am. With extra time before our 2 pm appointment at the school, we visited Deuk Chaa, a regional park with a zoo and fishery. The park abutted a beautiful field where farmers were collecting bananas using a horse-drawn cart and truck. For Chetra, Neary, and Toh, this was a nice opportunity to see some of the countryside, which they were not able to visit often from Phnom Penh. We then had lunch together in a special Kampong Cham picnic area, with a view of a lake. My three hosts decided to order frog, lake mussels, and barbequed lake birds, specialties of the area.

When we got to the first school, Chetra and Neary spoke to the director about what they were doing and asked for permission to take over some classes to advertise the application. School runs from 8 to 11 am and then from 2 to 5 pm. The students usually stay in school for only one of the two sessions. Chetra and Neary were granted permission to visit three classes, one per hour in the afternoon.

Our first class was a full room of forty-seven tenth graders. Chetra and Neary asked me to pass out stickers with the Khmer Rouge History application logo on them and a sign-up sheet. We opened a pop-up poster with the application logo on it. Chetra introduced herself and the team. "We are coming from Phnom Penh!" She started by asking about the Bophana Center—have they heard of it? None of the students indicated that they had, so she gave a brief description of what the organization does. She asked the students how much they know about Cambodian history. The students timidly answered some basic questions about the Khmer Rouge. Chetra explained the KR logo on the pop-up poster. She said that the letters

K and R are both figured in a way to represent people during the time of Pol Pot.

Then Chetra started to describe the Khmer Rouge history application. She asked who (of forty-seven students) had a smartphone; only one student pulled one out from his backpack. In this classroom, there was no power source for a projector, so Chetra stood in front of the room and showed how to use the application on an iPad. She explained, "You can read text, see video, listen to audio, and look at photographs." She showed different lessons from the application and explained why the audio and video are better than a simple book with the same information.

Neary set up a Wi-Fi hotspot and gave students the password and time to download the application from either the Play Store or the Apple Store. Neary told me that if the internet provided by the hotspot was slow, then they could download the application with their own data plan. Anticipating that not every student would have a smartphone, Neary and Chetra had brought twelve Samsung tablets so that the students could play with the application, which was already downloaded onto the tablets.

Chetra asked the students to press the button for lesson one—one student read the text out loud about Pol Pot. After she finished, Neary gave her a T-shirt (with the Khmer Rouge History application logo on the front). Chetra and Neary did not show videos or sound because they were worried that the internet might not be strong enough to support it. They later told me this had happened in a few schools and is a major challenge for the marketing trips. Approximately three students shared a phone or tablet. I noticed that some students who didn't have phones to share looked bored. Chetra later wondered out loud to me whether some students kept phones at home because they are not allowed in school.[36]

The next day, we went back to another branch of the same school in the morning and talked to a new set of three tenth-grade classrooms. Before going inside, we talked to the school director. He estimated for us that 40–50 percent of students have a smartphone. He said, pointing to a cell phone tower in sight, "We have fast 4G in this town—there is even a telecom tower next door." Chetra, Neary, and the director discussed their opinions on what might be the fastest telecom network in the province. The director said that teachers don't let kids use smartphones in school; if a teacher saw them using one, they would take it.

On the second day in the classrooms, the marketing routine was similar but we used a projector in addition to the iPad at the front of the class. All the projector equipment was brought from Phnom Penh—computer, screen, power strip, chargers for devices, extra batteries, speakers, and projector. Using the projector, Chetra went over the introduction to the Center and the Khmer Rouge history lesson with a PowerPoint presentation. Chetra again explained that the application had text, video, songs, and photos. This time they were able to play a video on the projector about the "Super Great Leap Forward," which they had downloaded beforehand.

On this trip, Chetra, Toh, and Neary brought to the provinces of Cambodia a product envisioned and produced by Bophana Center's team of writers, designers, and historians in Phnom Penh with the input of the REI Foundation and the EU aid organization. Like Srin in Oudong, they too savored the experience of being outside the capital and enjoyed what the provinces had to offer—food specialties and tourism sites. For the students they visited, they brought a new tool that carried with it historical media artifacts and a powerful interpretation of Cambodian history. In so doing, they connected these students more strongly to the smartphones, cell phone towers, Wi-Fi or 4G connections the app required.

These cases of new digital archiving and the provincial trips that they motivated deepen and extend concepts of the "enduring ephemeral," the imperial archive, and the politics of the platform. The endurance of ephemeral memories of the past happens in this context not just through online work but, crucially, through engagement with offline things and people, as I have illustrated in these descriptions. These provincial trips are primarily oriented toward collecting information (in the case of Amazing Cambodia) or toward spreading historical memory to new users (in the case of the Khmer Rouge History application). Yet they take on a deeper meaning for my interlocuters. During the wartime, provincial travel was difficult and dangerous. Srin's grandmother remembered fondly the car she owned before the Khmer Rouge era because it represented her former mobility; she told Srin how lucky he was that he and his generation could travel around the country. The collection and dissemination of digital memory importantly included eating special food, experiencing heavy rain, visiting tourism sites, and otherwise experiencing life in the countryside. The movement connected urban, emerging middle-class Cambodians to kinds of knowledge and people further removed from centers of global knowledge.

These cases also demonstrate how this offline work of maintaining the enduring ephemeral is obscured in the final products. Rendering this offline work invisible emerges from a legacy of the imperial archive, the colonial fantasy of consolidating authoritative knowledge, while systematically excluding some forms of place-specific and indigenous knowledge. Like the structure of the colonial archive that came before it, the format and norms of the online archives create conditions for what is authoritative knowledge and what is not. The logic of these digital tools and the way they are formatted necessitates the elimination of documentation of the "provincial trip": all that goes into the practices of collecting data (in Amazing Cambodia's case) and disseminating the app (in the Khmer Rouge app case) to the provinces. They do not capture a sense of the specificity and feeling of the place that the knowledge is used and disseminated in. But in both platforms, curating a new sense of historical memory also connects other parts of indigenous knowledge from marginal places to global centers of power. The platforms connect photos (of certain angles of the Cambodian countryside) to sites of information consolidation online. The sites are curated to create a certain image of Cambodia, most relevant to an imagined cosmopolitan audience.

That these two tools are technologies of global capitalism further ties them to the imperial archive's project. The use of contemporary online platforms for these archival projects solidifies ties between people and places (as of yet) lightly connected to centers of technology with emerging tools of globalized techno-capitalism. The use of Facebook gives the company license in perpetuity to the historical information that Srin posts. Facebook is problematic in its corporate nature in this case—cultural heritage is now newly commercialized through advertising. This case urges us to think through our ethical imperatives to build responsible technologies as well as responsible corporate and public sector policies that learn from the lessons of the history of colonialism and the Cold War to do better global work, which will be a theme of the conclusion.

The App-learning on Khmer Rouge History, developed by the Bophana Center with funding from the REI Foundation and the EU, uses not a corporate platform but instead an independently designed application, populated with high-quality films and sound materials, all with appropriate rights to access. The appeal of the application form is that it represents some "sexiness" for these foreign actors and foundations; there is excitement about reaching a new channel of users to spread the word about their commemoration and history learning. In so doing, however, the application opens up new problems in inequities in access to phones and data, and ties to telecom providers and smartphone companies. The creators found it difficult to encourage high school students to download this new application—data prices are expensive and Wi-Fi rare in much of rural Cambodia. The data-richness of the application leads to a higher quality application, but it is more expensive to download and run. Furthermore, many rural students didn't own smartphones and the Bophana application requires greater familiarity of and use of such digital tools.

These two projects perform infrastructural restitution in different, oppositional ways. The Bophana Center App-learning on Khmer Rouge History represents how foundations, nonprofits, and foreign actors, in conjunction with Cambodian practitioners, can develop a top-down form of media archive. Bophana Center has many monetary and cultural ties to European and Japanese influences; Rithy Panh himself is now a member of the diaspora and lives in Paris. The app's lessons remind Cambodian young people about the Khmer Rouge: the most violent and troubling aspects of the national history. This goal is in sharp opposition to the other kinds of infrastructural restitution that focus on the most inspiring parts of Cambodian history. Chetra, Neary, and their team were working from their heart. They faced challenges in attempting to connect to rural students, but they believed that the new tool was important and helpful for all students to promote learning about their country's past. They thought that knowledge about past violence could help prevent future violence. Their work is also an important corrective to Cambodian young people who do not know about the Khmer Rouge or who are exposed to Khmer Rouge–denying disinformation, which has blossomed on social media since 2017.

Amazing Cambodia, on the other hand, gives voice to a Cambodian researcher who represents the interests and curiosities about the past, largely romantic images of Cambodian 1960s modernism, which are shared by many of his urban middle-class peers (including the Roung Kon team and the Preah Sorya team). Srin used the tools of global capitalism but co-opts them for the purposes of restitution. Srin's Facebook page widely disseminates media remnants from the past that he and his audience view as scarce and important for construction of the next generation's collective memory and dreams of the future.

Many Cambodian young people do not feel safe publicly posting statements against the ruling party or calling for change. Instead of doing so directly, they instead use the praise of historical moments as a way to call for social change with subtlety. So, Srin here—instead of saying explicitly that he wanted public support for the arts—praises the 1964 period when Cambodia had public support of the arts. The same goes for other instances on his page, which calls for more urban green space, more public housing, and better urban planning through infrastructural restitution. He enmeshes commemoration (of artists, architects, creatives, intellectuals, and a way of life lost) with ideals of the past as visions for the future. In the modernist past, many Cambodian young people see possibility: one with public housing, a rich creative sector, and government investment in national infrastructure.

CONCLUSION

I have insisted in this book on thinking about technology with a forward and backward vision: cultural memories—including histories of conflict and artistic heritage—inform future visioning, including technology design and appropriation. In Cambodia, historical media modes and contemporary ones coexist. Media ruins populate the streets of Phnom Penh. This coexistence of new and old media infrastructures makes material the crossing of memories, skills, and people across historical and contemporary moments. Every place lives with the shadow of its history; however, in Cambodia, the events of recent history are present in contemporary life in a striking way. Recognizing this force of history is essential for any technology implementation effort. The concept of infrastructural restitution—the creative reconstruction of historical media artifacts and infrastructures—illustrates ways that history matters to vernacular innovation, including in the realm of emerging technology.

Each of the chapters illuminated a different aspect of infrastructural restitution. The first chapter introduced the value of an infrastructural lens to understanding an intertwined history of media and politics. Chapter 2 first presented empirically the concept of infrastructural restitution, pointing to the repair work of state media workers reconstructing infrastructure after disaster. Chapter 3 noted dangers in ahistorical foreign media intervention. The second half of the book illuminated the ways that media preservation projects offer a forward-looking glimpse toward post-conflict healing and political action. Chapter 4 elucidated the hidden script of infrastructural restitution, and the ways that it can encourage issue-level political change. Chapter 5 explored the ways that excavating positive affect from historical media artifacts can be a mode of healing. It showed how processing history materially and watching, rewatching, and interpreting media artifacts can shift emotional experience. Chapter 6 illustrated the transnational tensions of presenting past artifacts in the contemporary context, and the power of new media archives in telling contested histories. It also explored what kinds of restitutive work online contexts render invisible.

The three historical chapters were not just background to the main story of contemporary digitization. Rather, the first half of the book painted a concrete picture about the historically tight links between media and politics. Urban dwellers in Phnom Penh live and make meaning among media ruins from each of these earlier periods. The infrastructures and artifacts built in the first chapter are what reconstructors now primarily seek to restore. The links between the past and present are not just material and architectural, though; some of the actors working during the PRK period are still active in the media sector. In addition, many of the current leaders of Cambodia are veterans of its conflicts. Their personal histories and the ways they acted in specific ways in their communities matter today to the ways that they govern and intimidate citizens. As such, each participant's interpretation of history reflects their current positioning and desires for the future. These chapters also heed a warning about the necessity of paying attention to history in any technology intervention. As I showed particularly in chapter 3, ahistorical approaches to media can cause damage and instigate violence.

An infrastructural approach offers new insight into the practices of restitution, as well as its reach and meaning, with its interlocking interest in form, work, and relationality. In Cambodia, form and content are inseparable dimensions of media reconstruction. Much of the contemporary media infrastructure was installed by prior regimes and forces encounters with the past. Paying attention to the spaces, films, and platforms in and through which such reconstruction happens shows how memory manifests in cracks, breaks, and crumbling buildings of media, not just through stories on screen, old photographs, or recovered radio reels. The repair work of young media creators happens in different formats: the physical space of media (cinemas), the materiality of media (film), and the storage of media (internet "archives"). In various ways, all of these projects move across online and offline realms, digital and analog formats, and in architectural space. I considered form in both its material (radio transmitters and receivers, the number and location of film projectors, fiber-optic cables) and immaterial instantiations (networks, memories, ghosts, and hauntings).

CONCLUSION

The infrastructural *work* of restitution requires skill, cooperation, and attention, and can be highly affective. The work also often involves care and collaboration; for example, the Roung Kon team worked together to measure the distance between door frames in old cinemas, with older and more experienced volunteers teaching the younger ones to enter the data into AutoCAD. This cooperative and helpful work of infrastructural restitution helps contribute to its effects for peace-building and reconciliation. It can also involve difficult emotions, like fear and sadness; for instance, Ka Toy traveled through the northwestern provinces as a projectionist in the 1980s, facing attacks from warring factions, in order to screen films for rural populations. As a form of infrastructural work, building from scholars like Star and Strauss, this historical and healing work is often invisible.[1] In particular, I showed in chapter 6 how some of the offline, community-embedded work, like visiting provinces for tourism, collecting new stories, and teaching students, is rendered invisible to globalized audiences of internet tools.

Paying attention to this invisible work brings me to the underlying feminist motivation of this book—in a world in which much is broken and in a particular context too often stereotyped for its national tragedy, I focused in this book on healing, collaborative, and inspiring action. The work of infrastructural restitution is rooted in the past, but it exists in the present and pushes toward the future. I described the development of Preah Sorya's large volunteer networks, Meta Moeung giving up personal time and space for the generation of a new invisible infrastructure of support for young artists, and Sokmean Srin spending his nights and weekends learning about the best parts of Cambodia's past so he could teach others about it. All of these are purposeful and generous activities that, regardless of their outcomes, are leading to powerful collectives and meaningful work that improve people's lives.

The *relational* aspect of media infrastructures and their inherent power dynamics are clear in the links between national politics, foreign interference, and technology in Cambodia. Media infrastructures drive and/or enable domestic politics and their relation to global empires. Understanding the values of infrastructural restitution cannot be understood without political context, rendered in historical perspective. In the first part of the book, I demonstrated the ways that domestic and foreign politics hinged on the use of media technologies during three significant moments in the Cold War. In the first chapter, I argued that the United States Information Service supplied tools and training for a media infrastructure that Sihanouk then used to undergird his authoritarian state. This became clear, for example, when Sihanouk cut off ties with the United States after discovering that they provided radios to Dap Chhuon's right-wing rebels. The second chapter described the ways that Vietnam and the Soviet Union sponsored media training and sent devices to Cambodia to support state power and post–Khmer Rouge recovery during the People's Republic of Kampuchea period (1979–1991). In the third chapter, I told of the development of Radio UNTAC, which was a key part of the first national election in 1993 and the opening of Cambodia to Western markets. I then showed that, after the opening of markets during the UNTAC period, multinational telecom companies dictated the building of infrastructure and the labor conditions of such infrastructures (for instance, those who dug the ditches for the first fiber-optic cable to run through Cambodia).

In the second half of the book, I described projects of infrastructural restitution in a moment of renewed media authoritarianism. In each case, I showed how young media creators, who did not themselves live through the Khmer Rouge regime, have called for political and social change through historical reckoning. The cases highlighted different dimensions of future-building, from Roung Kon's appeals for urban public spaces and space for expression in a rapidly urbanizing, gentrifying Phnom Penh to Preah Sorya's moves toward emotional healing to Sokmean Srin's romantic calls for the cultural flourishing of the past. The Bophana Center, through its Khmer Rouge History application, gave a case for a future in which all young people have a clear sense of the past. In their different ways, all of these projects used media as prompts to remember, to forget, and to use memory to move through a collective history of violence.

Whether working in decaying cinemas or at ancient royal temples, in an arts center or a school, all of these participants used historical artifacts—mediated through material platforms—to build new futures and call for change. As political actions, practices of infrastructural restitution tend to be subtler and more targeted to specific issues than more publicly obvious forms of protest, including the networked protest we have seen from the Arab Spring to Hong Kong.[2] In a moment of increased authoritarianism, however, the subtler forms of infrastructural restitution may be more palatable and effective as a form of political action in Cambodia.

CONCLUSION

That is not to say that this political action is uncontested; on the contrary, in each case, actors with different agendas could articulate counter-arguments or alternative interpretations of the historical artifacts that are recovered and celebrated by these groups. Restitution brings us back to a previous state, but we can imagine ambivalences about that return. For instance, romanticizing the Sangkum Reastr Niyum period can be contested on the grounds that it represented a time of deep inequality, as illustrated in the first chapter of this book. When Roung Kon visited the Hemakcheat cinema, people were living there in slum-like conditions. The group praised the cinema for its past cultural glory without directly attending to contemporary urban inequality, which mirrors the inequality of the time that they are honoring. These impulses can work alongside nationalist or lineage-based impulses that praise the past uncritically. They can also lead to xenophobic harms. As discussed in chapter 2, encouraging change in ahistorical ways, the way that UNTAC did, however, is not an anecdote to the romanticization of the past as it can also encourage more violence.

Bringing historical media artifacts back to life can also unearth some uncomfortable realities about the past for younger generations. For instance, the participants of "Listening from the Archives" described disorientation in hearing digitized communist propaganda from a 1984 radio program. Just as the law of restitution makes things return to a state as if the conflict never happened, infrastructural restitution cannot fix what happened in the past. Rather, these practices honor the artists who died and the cultural artifacts that were lost or forgotten in the later wars, and encourage the consideration of how the best of history can be integrated into the future.

Media infrastructure and its repair cannot be understood solely on the political register; they must also be understood on a personal and affective register.[3] A collective history of violence has impacted the political, cultural, and technological course of Cambodia since the late 1960s. Artists are attuned to the emotional and embodied legacy of that violence. From the earliest parts of this story, the USIS imported film and radio technology for the purposes of ideological control. These tools were taken up by Sihanouk's authoritarian state. We see the affective and political tangled up in media in immaterial instantiations like Sihanouk's fight to control radio waves, or in material things like Cold War–legacy radio transmitters. Independent artists used these same tools to make funny, touching, silly, or otherwise affectively evocative media content.

The concept of disintegration noise brings together the material and psychological aspects of a history of violence through material form. The wear of time is visible in the marks and scratches of old film reels, the haphazard copying of these reels into VHS formats, and marks of the program used for their digitization. Watching the limited number of films available from the prewar period again and again mirrors the repetition common to ruminating over a painful history. The effect for Preah Sorya is to feel all the feelings of the past, a complex mix of loss and the happiness of the images on screen, and thus to shift the "problems of the way of the heart." Recovering media gives space for positive affective experience of the past.

Media artists' relation to violence and memory changes the meanings they give to repair, and these modes of restitution provide new insights into its theorization.[4] In a post-conflict setting with a broad aesthetic of destruction, noise and brokenness can act as complex structures of feeling and indicators of value rather than problems to be fixed. The horizontal lines of a bad VHS copy of a film, now reproduced in a digital copy, convey authenticity for Preah Sorya. A vocal performance can create something new from that which has been lost. When Roung Kon praises the media ruin and mourns its demolition and replacement with new buildings, we can see how the choice not to repair can itself be a value-laden act. These cases pointed to the ways that both decay and ruin, as well as repair and reconstruction, import culturally specific meaning and can differ based on any individual's relation to history.

By drawing lines between the historical media landscape and contemporary Cambodia, I illustrated the Cold War roots of the ways that digital media technologies today have broadly become tools for global geopolitical interference, nationalism, and authoritarianism. This book therefore provides a historical foundation for understanding the contemporary geopolitics of technology in Cambodia: the ways that digital media policy and practice are again becoming increasingly central to internal politics and foreign relations. Across the Southeast Asian region, authoritarian regimes govern through the internet, making social media a part of civil service and a site of resistance.[5] One particularly relevant issue across the region in 2022 is the expansion of Chinese influence in matters of digital media policy. Within Cambodian online spaces, there is rising tension between Western corporate technology policies and Chinese state technology practices. Government actors and citizens sometimes take different sides on these geopolitical media tensions.

CONCLUSION

For example, the Cambodian government's digital media advisors and allies are more frequently now coming from China. In February 2022, the government proposed running all internet traffic through a National Internet Gateway, allowing authorities to monitor all activity and collect users' data, a policy similar to Chinese government internet surveillance.[6] At the same time, Facebook remains by far the most popular way for most Cambodians to use the internet. This semi-monopolistic entity uses corporate surveillance and extract advertising revenue from one poorer part of the world to a richer one.[7] Through the case of Amazing Cambodia, I addressed the postcolonial politics of platforms and raised a concern about the power of a corporate American transnational technology platform (Facebook), which now holds license to valuable historical cultural heritage material about Cambodia. The US government is now using such Chinese global competition as a way to encourage more funding to US tech firms so they can have a broader international reach, thereby winning the capitalist rivalry with China.

As mentioned in the introduction, my intention here is not to mark Chinese or American approaches to digital diplomacy as wholly different; rather, they have different modes of control and exploitation building on long-standing modes of foreign intervention via media. Both demonstrate strong connections between the state and technology companies. For instance, the links between Facebook and the US state are multiple. Facebook spent nearly $20 million lobbying in the US in 2020, financially connecting the company to the state. The company has also made important decisions related to advertising and speech that directly impacted political power and swayed democratic elections. Likewise, the major tech platforms in China support state surveillance and do not allow for oppositional speech. The socialist state and the corporate sector overlap strongly such that China's economic model has been called "capitalism with Chinese characteristics."[8]

Instead of encouraging the continuation of polarized language, my goal is to show that both digital governance models have troubling concerns for postcolonial politics when enacted in sites like Cambodia. Digital media diplomacy is affecting amazingly dynamic media infrastructures, including geographies of technology installations (such as 5G towers and the location of new data centers) and information control tactics (such as chilling effects and platform censorship policy). Analyzing these geopolitical tensions in conjunction with the histories of media and politics presented in the first half of this book illuminates that these new geopolitical trends are far from exceptional and part of an ongoing legacy of domestic and foreign impulses for control.

Many scholars have responded to the problems inherent to transnational digital tools and policies by calling for changes to the tools themselves through corporate self-regulation. In the field of human-computer interaction, the "implications for design" section is a common area in which to call for platform change. Though these are often apolitical changes at the level of hardware or software, some more meaningful improvements that scholars have suggested—and, at times, companies have enacted—include enhanced content moderation, more contextually attuned speech policies, and proper translation of user interfaces.[9] Though we can encourage responsible self-regulation, there are limits to this self-monitored approach. Given the history raised here and the essentially extractive business model of corporate technology platforms, self-regulation is not a radical solution, and likely not a wholly effective one.

As an American and former tech worker, I feel some responsibility for reining in the harms of American companies in international settings through regulatory channels, particularly given what I have described about the violence that America has historically enacted through media channels (via the USIS, for example). US-based users of Facebook could call for US legislation that would give rules for the company's moderation of international speech, minimum bars for user interface translation and the development of digital literacy, and consent tools that are actually understandable. Other possible avenues for change are through international regulation. For instance, the UN could regulate transnational technology platforms based on international human rights law. Given the transnational nature of the internet and many internet tools, having internationalized laws for antitrust, privacy, and speech would give possibility to more just global connection via the internet. Using international or US legislation to regulate the use of platforms in other nations, however, is complicated by conflicting desires for national sovereignty and self-determination and the challenges of international enforcement.

The practices of infrastructural restitution I have described shed light on broader questions of transnational technology use. Much of the work described in this book uses hardware and software in platforms developed and designed outside Cambodia, but media creators made them their own and embedded them within their own political practices and historical context. These, like other globalized objects, are rendered local and situated as a form of global assemblage.[10] The actors in this book have appropriated these tools for their own uses, reconfiguring lines of power and agency in the global technology landscape. The creative tension is generative for remembering something old while building something new, and gives users an important place in the iterative and cumulative process of technology evolution. This interpretation builds on critiques of unidimensional design for others without their input, through either an analysis along the lines of power and economy, such as postcolonial computing, or the praise of flexible and fluid technologies.[11] It also reiterates the values of seeing repair as a mode of artfulness, environmentalism, and healing.[12]

I cannot resolve from this narrative the essential tensions between the controlling nature of these platforms and the tailored and grassroots uses of the tools I have described. This tension has long been a feature of Cambodian media use. For instance, Som Sam Al's training through the USIS and with Sihanouk prepared him for making independent films with artistry. We see the same phenomenon happening today with artists or media creators like Sokmean Srin co-opting tools of control like Facebook for their own political and cathartic ends. This book gives more evidence of ways that the technology industry—from tech platforms like Facebook to data centers and smartphones—represents a large, important, and relatively new transnational influence in many people's lives around the world. The troubling political economy of the platforms does not preclude their use for artistic or activist purposes, but presents a barrier that their users must overcome or overlook.

My hope for the future is that artists can use locally generated platforms for grassroots causes, written in local languages and customized for local infrastructures, through nonexploitative business models. I hope that there become more opportunities to host healing, creative, and socially engaged content on tools that fully align with such content. The realization of this vision, however, likely requires seeking out alternatives to the dominant political economy of platforms we know today and a rethinking of the contemporary global internet environment in ways that are safe, affect-sensitive, and in tune with global power dynamics.

NOTES

INTRODUCTION

1. Mekong Strategic Partners and Raintree, "Cambodian Startup Ecosystem Report 2018," 3.

2. Vong and Hok, "Facebooking: Youth's Everyday Politics in Cambodia."

3. Miller, *Tales from Facebook*.

4. For more on Facebook, politics, and disinformation, see Vong and Sinpeng, "Cambodia."

5. For more sources on twentieth-century Cambodian history, see Becker, *When the War Was Over*; Chandler, *The Tragedy of Cambodian History*; Strangio, *Hun Sen's Cambodia*.

6. The Extraordinary Chambers of the Court of Cambodia found Khieu Samphan and Nuon Chea guilty of genocide against Vietnamese and Cham peoples and "crimes against humanity" against other Khmer people in 2019.

7. Pirozzi, *Don't Think I've Forgotten*; Chou, *Golden Slumbers*; Daravuth and Muan, *Cultures of Independence*.

8. Thompson and Murphy, "Cambodia Is Turning the Tide on Looted Statues, but Some Things Cannot Be Returned."

9. For more on "Complexifying Restitution," see its description from the Berlin Biennale: https://12.berlinbiennale.de/artists/jihan-el-tahri/.

10. On memory as contested and historically contingent, see Jelin, *State Repression*.

11. Parks and Starosielski, *Signal Traffic*; Dourish, *The Stuff of Bits*.

12. Hu, *Prehistory of the Cloud*.

13. Parks and Starosielski, *Signal Traffic*; Mattern, "Scaffolding, Hard and Soft."

14. Parks and Starosielski, *Signal Traffic*; Larkin, "Politics and Poetics of Infrastructure."

15. Star and Strauss, "Layers of Silence, Arenas of Voice."

16. Star and Ruhleder, "Steps toward an Ecology of Infrastructure."

17. Hackett et al., *Handbook of Science and Technology Studies*.

18. Pipek and Wulf, "Infrastructuring"; Jack, Chen, and Jackson, "Infrastructure as Creative Action."

19. Nguyen, "Infrastructural Action in Vietnam."

20. Star, "The Ethnography of Infrastructure."

21. Star and Bowker, "Enacting Silence."

22. Star and Strauss, "Layers of Silence, Arenas of Voice."

23. Hochschild, "Foreword: Invisible Labor."

24. Poster, "Who's on the Line?"

25. Lindtner, *Prototype Nation*.

26. This insight is articulated clearly in Star and Ruhleder, "Steps toward an Ecology of Infrastructure."

27. Star, "Ethnography of Infrastructure."

28. For example, Anand shows the ways that poor people in Mumbai have to join (physical) water infrastructure with social practices of "pressure" to municipal governments to get water. Anand thus points to Mumbai's water supply as a social *and* physical system. Anand, "Pressure."

29. Richards, *Imperial Archive*; Stoler, *Along the Archival Grain*.

30. Postcolonial memory, as Kusno describes, "is a fraught terrain, contestory and multistranded, and woven around the politics of inclusion and exclusion, of remembering and forgetting." Kusno, *After the New Order*, 12.

31. Gordon, *Ghostly Matters*.

32. Edwards, *Cambodge*.

33. Edwards, *Cambodge*.

34. Ly, *Traces of Trauma*, 2.

35. Edwards, *Cambodge*.

36. Updegraff, Silver, and Holman, "Searching for and Finding Meaning in Collective Trauma."

37. Coombes, *History after Apartheid*; Kwon, *After the Massacre*; Nelson, *Reckoning*; Olsen, *Tailoring Truth*.

NOTES

38. Schwenkel, "Recombinant History."

39. Schwenkel, *The American War in Contemporary Vietnam*.

40. Jack and Avle, "A Feminist Geopolitics of Technology."

41. Elwood, "Digital Geographies, Feminist Relationality, Black and Queer Code Studies."

42. Tsing, *The Mushroom at the End of the World*; Escobar, *Designs for the Pluriverse*.

43. Taking inspiration, in part, from Mrázek, *Engineers of Happy Land*.

44. Wolford, "The Difference Ethnography Can Make."

45. I have tried in my writing and presentation to move away from the language of "fieldwork," which still relies on a legacy that believes that some places are centers of research and intellectual thought, whereas other places are sites to be researched—with the further assumption that the knowledge flows back "from the field" to the centers of knowledge and research. For more, see Gupta and Ferguson, "Discipline and Practice."

46. Burrell, "The Field Site as a Network."

47. Burrell, "The Field Site as a Network."

48. For instance, the use of Facebook in Myanmar, during both the Rohingya displacement and the 2021 coup, was covered actively by the *New York Times* and other American press outlets.

49. This was a paid position; I paid my RAs $15 per hour or $50 per day.

50. I used the Non-Standard Romanization System by Wanna Net, which can be found at https://ethnomed.org/wp-content/uploads/2020/02/Non-standard-Romanization-system1.pdf. Thank you to Catriona Miller for her editorial help with this.

51. Willis and Trondman, "Manifesto for Ethnography."

52. Emerson, Fretz, and Shaw, *Writing Ethnographic Fieldnotes*; Wolfinger, "On Writing Fieldnotes."

53. Strauss, *Qualitative Analysis for Social Scientists*; Strauss and Corbin, *Basics of Qualitative Resarch*; Strauss, *Continual Permutations of Action*.

54. Muan, "Citing Angkor."

CHAPTER 1

1. The film was finished in 1979 while Sihanouk was living in North Korea.

2. These three dimensions correspond to the three insights from infrastructure studies that I highlighted in the introduction.

3. I am in part inspired here by the field of media archaeology. Media archaeology insists that historical media artifacts can teach us about media's past (in addition to documents and content). For an introduction, see Parikka, *What Is Media Archaeology?*

4. Sihanouk's power to shape information about him was impressive, and a great deal of the archival record is tinged with propaganda created by Sihanouk himself—therefore making it sometimes difficult to assess Sihanouk's popular support. Sihanouk's party was in charge of many of the major magazines of the time, including *Réalités* (in French) and *Kambuja* (in Khmer). Despite the tension and blights on the shininess of the so-called golden era, Sihanouk denied any reports that suggested there was tumult in Cambodia.

5. Roger Nelson told me that Ingrid Muan's papers were at the National Archives, and Srin Sokmean told me about a few PRK-era documents in the National Archives.

6. Erich DeWald shows that the French did construct a shortwave radio network under the name "Radio Colonial" in 1931, an empire-wide limited relay network transmitted from Lyon that inconsistently reached colonies. French people living in the capital also created a small hobbyist private club for radio in the 1930s called Radio Phnom Penh. The most popular brand of radio in Indochina was Philips (a Dutch brand), but since the French placed a yearly quota on import licenses and a 125 percent ad valorem tariff on extra-imperial trade for non-French goods, radio purchases were expensive and sales were a poor business. DeWald, "Taking to the Waves."

7. DeCoux put radio direction under Jacques Le Bourgeouis. See "L'amiral Decoux et Quatre Ans de Radio," August 1, 1944, *Radio Indochine*, National Archives of Cambodia.

8. *Prakas* from 1939 and 1941, National Archives of Cambodia.

9. The journal *Radio Indochine* complained about the technical difficulties of covering some Cambodian topics due to distance and technical capacity. They also had trouble staffing for Vietnamese and Khmer cultures; in 1945, they attempted to incorporate Cambodians into a planned advisory group to improve the radio, though there is no evidence of this advisory group ever materialized. See "Futurs problemes de l'organisation de la radiodiffusion en Indochine," February 1, 1945, *Radio Indochine*, National Archives of Cambodia.

10. Chandler, *The Tragedy of Cambodian History*.

11. In 1946, they invested in a "mobile recording studio-bus" hoping that "radio reporting will take on an importance it has rarely had since the invention of radio." See Jacques Sallebert, "Radio Saigon, the Voice of France in the Pacific," February 22, 1946, no. 70, *Radio 46*, National Archives of Cambodia.

12. Donald Heath to Washington, "Topic: Radio as a Vehicle for USIE Purposes in Indochina," July 22, 1950, Ingrid Muan Papers.

13. Mitterand, *Norodom Sihanouk, King and Filmmaker*.

14. Steinberg, *Cambodia: Its People, Its Society, Its Culture*. This text, from 1957, is a part of the "Country Survey Series of the Human Relations Area Files," a nonprofit research outfit affiliated with Yale created "to provide an interpretative, integrated description of selected societies in Europe, the Middle East, Africa, South and Southeast Asia and the Far East." Steinberg's descriptions are often rather superficial and his data collection method is unclear; however, the book contains a survey of the media available in Cambodia at different points of the 1950s.

15. Steinberg, *Cambodia: Its People, Its Society, Its Culture*.

16. Steinberg, *Cambodia: Its People, Its Society, Its Culture*.

17. Ingawanij, "Itinerant Cinematic Practices."

18. Chandler, *The Tragedy of Cambodian History*, 189.

19. The USIS was active in many parts of the Third World, and media were a crucial component of the wider Cold War. In Eastern Europe, for example, the USIS was deeply involved in the establishment and management of Radio Free Europe, which was said to be a voice of the anticommunist opposition. For background on Radio Free Europe, see Granville, "Radio Free Europe." The USIS funded mobile cinema, particularly public health films, in locales as diverse as Nigeria and Thailand, as discussed in related scholarship by Brian Larkin and May Adadol Ingawanij. For more, see Larkin, *Signal and Noise*; Ingawanij, "Itinerant Cinematic Practices." Sometimes American influence was packaged into secretive CIA activity. The anticommunist and pro-capitalism and democracy cause, for example, led the American CIA-backed agency the Asia Foundation to fund the production of anticommunist films in Korea. For more, see Lee, "Creating an Anti-Communist Motion Picture Producers' Network in Asia." The development of media for politics was not limited to US activities, either; elsewhere in Southeast Asia, colonial powers were trying to control the same audiovisual channels. In colonial Singapore and Malaya during a similar period (1948–1961), a number of British governmental institutions invested heavily in British anticommunist propaganda filmmaking. See Aitken, "British Governmental Institutions."

20. For more on film activities in South Vietnam, see Rouse, "South Vietnam's Film Legacy."

21. The text reads: "the time now seems ripe to press for opportunities to increase US collaboration into the cultural and particularly the radio field. The latter medium is probably the most important now available for acquainting the Vietnamese with our purposes and challenging the Communist lies. So far, we are chiefly represented

on the radio by French interposition. Moreover, French propaganda here has largely fumbled its opportunities. On the basis of the American interests at stake, we need to see effective propaganda use made of the radio stations. An American with radio program and technical ability is needed to give continuous guidance and encouragement to the stations of the local authorities, and to take an effective interest in helping them acquire needed replacement parts and new equipment. Such a man would keep watch over the appeal of the programs and the technical quality of the transmission. He would also help to sustain the courage of the station managers." See Donald Heath to Washington, "Topic: Radio as a Vehicle for USIE Purposes in Indochina," July 22, 1950, Ingrid Muan Papers.

22. RG469 Records of Foreign Assistance Agencies 1948–61 Mission to Cambodia, Office of the Director, Jules Suby 1952–56, Box 14, March, 1956, "1956 Report on Mass Communications in Cambodia," Ingrid Muan Papers.

23. USIS telegram November 14, 1950, Ingrid Muan Papers.

24. The text reads: "The objection arises whenever Cambodia's individuality is submerged and its entity as a distinct kingdom with its own history, language and culture is lost sight of." USIS Telegram from Saigon 133 to Department of State, September 7, 1950, Ingrid Muan Papers.

25. Chandler, *The Tragedy of Cambodian History*, 189.

26. Osborne, *Sihanouk: Prince of Light, Prince of Darkness*.

27. Of these radios, 285 came from the ICA (an American aid organization) and 1,000 receivers from the Colombo plan. The Colombo plan was an Australian initiative to stop communism in Asia. In March 1955 the DEA enlisted E. S. Heffer, a radio engineer for Amalgamated Wireless Australasia, to conduct a six-week technical survey of the capacity of Indo-China and Thailand to use and maintain portable radio receivers, and they determined that they would distribute 1,000 sets in Indochina with South Vietnam as the highest priority. At the request of the South Vietnamese Minister for Defense, Australia used Colombo Plan funds to supply military units with petrol-driven generators for use in wireless transmission. Based on the notes from the US Foreign Service and USIS, some of these sets were clearly also distributed in Cambodia by the USIS on behalf of Colombo powers. See Oakman, *Facing Asia: A History of the Colombo Plan*.

28. A USIS Phnom Penh note from November 1956 states that monks particularly "appreciate[d] the prestige of listening and then playing the 'oracle' to the populace." Ingrid Muan Papers.

29. A number of documents in the Ingrid Muan Papers report ways the USIS trained technicians and repaired earlier equipment. RG469 Records of Foreign Assistance Agencies 1948–61 Mission to Cambodia, Office of the Director, Suby Jules 1952–56, Box 14, March 1956, "1956 Report on Mass Communications in Cambodia." In

addition, two United States Operation Mission (USOM) reports from 1955 and 1956 describe a lack of technical capacity in the Cambodian Ministry of Information to use the products they had supplied earlier in the decade: USOM/Phnom Penh, Subject: "The Report of the Program Support Division," August 19, 1955, Ingrid Muan Papers; RG469 Records of Foreign Assistance Agencies 1948–61 Mission to Cambodia, Office of the Director, Suby Jules 1952–56, Box 14, March 1956, "1956 Report on Mass Communications in Cambodia," Ingrid Muan Papers.

30. USOM/Phnom Penh, Subject: "The Report of the Program Support Division."

31. The USOS published a Khmer-English magazine in 1957 that reported the start of Voice of America in Cambodia. Ingrid Muan Papers.

32. This is a photocopy of a report held in Ingrid Muan's papers. General Records of the Department of State 551.51h. 1950–1963 Central Decimal File Box 2156, "Education Exchange Report and Joint USIS Message," May 28, 1958; August 12, 1955, "Dept of State to AmEmbassy Phnom Penh, Cablegram from Pres Eisenhower to King Norodom Suramarit to be delivered August 15th"; August 15, 1955, Telegram #219 "McClintock to Secretary of State, Norodom Sumarit to President Eisenhower," Ingrid Muan Papers.

33. Telegram #199 McCintock to Secretary of State, "Matters related to USIA," August 15, 1955, Ingrid Muan Papers.

34. "Project requirements for 1956," August 19, 1955, Ingrid Muan Papers.

35. RG 306 Records of the US Information Agency Office of Research/ Field Research Reports 1953–1993, Box 15, Cambodia 250/67/06/01–07, Dispatch 60 USIS Phnom Penh to USIA Washington DC, Subject: "VOA Listeners' Response," May 3, 1957, Ingrid Muan Papers.

36. Walling of USOM PP to ICA Washington National Museum Cambodia, February 1956, Ingrid Muan Papers.

37. Telegram from Wilder to Secretary of State of Information, August 14, 1957, Ingrid Muan Papers.

38. For another perspective on USIS mobile cinema in Cambodia, see Muan, "Playing with Powers." This article was published posthumously.

39. Though villagers may not have had a choice whether or not they would go to see a USIS film when a cinecar or cineboat showed up in their town.

40. Walling of USOM PP to ICA Washington, February 1956, Ingrid Muan Papers.

41. Documents on these trainings in Ingrid Muan Papers. See also Muan's dissertation, "Citing Angkor: The 'Cambodian Arts' in the Age of Restoration, 1918–2000," which has an image of a training course on the "operation, repair, and maintenance of film projectors sponsored by US aid for the Cinecar project 1957" on page 229.

42. 1956 Foreign Service Dispatch, Subject: "Film Showing by USIS Boat in Battambang," USIS/Phnom Penh, November 6, 1956, Ingrid Muan Papers.

43. USIS Phnom Penh note from November 1956, Ingrid Muan Papers.

44. I/S I/R USIS Phnom Penh from October 1956, Ingrid Muan Papers.

45. "Report from Cultural Centers," December 1956. They note that there was a total audience of 17,420 in Battambang alone. Ingrid Muan Papers.

46. RG 469 Records of the US Foreign Assistance Agencies 1948–61 Mission to Cambodia, Office of the Director, Subject Files 1952–58, Box 7–17 1957, Updated Battambang file, 1957, Ingrid Muan Papers.

47. As Brian Larkin notes, this style of USIS films (particularly those on public health) can be distinguished from other kinds of cinema because they are fundamentally noncommercial and therefore create a different subjective relation between film and viewer than films traditionally sold in cinemas. Larkin, *Signal and Noise*.

48. John M Anspacher, Country Public Affairs Officer, Records of US Foreign Assistance Agencies, 1948–61, Office of Far Eastern Operations, Cambodia Subject Files 1955–61 Boxes 14–32 1957–59 250/75/34–35/07–01, Photo (film) Box 14, January 3, 1957, Despatch 36 USIS PP to USIA Washington, Hilites December 1956, Ingrid Muan Papers.

49. 1956 Foreign Service Dispatch, USIS/Phnom Penh, November 6, 1956, Ingrid Muan Papers.

50. Memo from September 22, 1957, Ingrid Muan Papers.

51. Steinberg, *Cambodia: Its People, Its Society, Its Culture*.

52. Steinberg, *Cambodia: Its People, Its Society, Its Culture*.

53. Steinberg, *Cambodia: Its People, Its Society, Its Culture*.

54. Ebihara was the first American anthropologist to conduct ethnographic research in Cambodia and the last for three decades.

55. Ebihara, *Svay: A Khmer Village in Cambodia*.

56. Steinberg, *Cambodia: Its People, Its Society, Its Culture*.

57. There is a point of contention between Ebihara and Steinberg about the role of the pagoda in disseminating radio broadcasts. May Ebihara explains that Steinberg emphasized the pagoda as a key point in the relay of information and in opinion formation. She rebutted: "This may well be true in some regions and for isolated villages . . . but this was not the case in Svay. While several monks at the local temples are impressively intelligent and well-informed, they do not appear to be significant sources of secular news and information for the community. Neither

do the Svay villagers make a point of congregating at the temples to read books and papers or listen to the radio as Steinberg suggests." Steinberg says, "The more important pagods and the quarters of senior bonzes usually have radio sets. The receivers are likely to be powerful and far-ranging because the bonzes like to listen to the Buddhist programs that are broadcast by stations in Burma and Thailand." Given these differing observations, there were likely a range of uses of radio within any given pagoda depending on the politics and Buddhist sect of the temple.

58. Steinberg, *Cambodia: Its People, Its Society, Its Culture*.

59. Rust, *Eisenhower and Cambodia*; Crane, "Red-Handed: Pinning the Blame for Dap Chhuon on the CIA."

60. Rust, *Eisenhower and Cambodia*.

61. "Communist Media Developments—IRI/PI Briefing Notes," November 4, 1957, Ingrid Muan Papers.

62. Steinberg, *Cambodia: Its People, Its Society, Its Culture*.

63. "Communist Media Developments—IRI/PI Briefing Notes."

64. Radio Free Asia, "China-Cambodia Relations: A History Part One."

65. "Allocation de M Chau Seng Ministre de l'information à l'occasion de la ceremonie d'inauguration de la Station Royale de Radiodiffusion Khmère de Stung Meanchey," May 6, 1960, National Archives of Cambodia.

66. "Allocation de M Chau Seng Ministre de l'information."

67. "Allocation de M Chau Seng Ministre de l'information"; the earlier transmission strength comes from Steinberg, *Cambodia: Its People, Its Society, Its Culture*.

68. "Alloculations de LL EE Pho Proeung Haut Representant de SAR le Prince Norodom Sihanouk, Chef de l'Etat, et Chen Shu-Liang, Haut Representant de S. Exc M Chou-en-Lai, Premier Ministre du Conseil des Affaires d'Etat à l'occasion de la ceremonie de remise de la 2eme tranche des travaux de construction de la Station Royale de Radiodiffusion Nationale Khmère," June 21, 1962, National Archives of Cambodia.

69. Airgram G-278 Trimble, AmEmbassy PP to Sec of State, 1961, Ingrid Muan Papers.

70. Leos, "Economic Ties with U.S. Cut in 1963."

71. Chandler, *The Tragedy of Cambodian History*.

72. "Lese Majeste: The Khmer-Serei Radio," November 24, 1964, M-675–64, Ingrid Muan Papers.

73. For more on the 1965 dissolution of ties, see Leos, "Economic Ties with U.S. Cut in 1963"; Bong, "Cambodia's Disastrous Dependence on China."

74. USIS notes from June 11, 1957, Ingrid Muan Papers.

75. *Réalités*, January 12, 1967, National Archives of Cambodia.

76. *Réalités*, January 27, 1967, National Archives of Cambodia.

77. *Réalités*, January 27, 1967. The frequencies are "195 meters" and "405 meters in middle waves." National Archives of Cambodia.

78. *Réalités*, February 10, 1967, National Archives of Cambodia.

79. *Réalités*, February 24, 1967, National Archives of Cambodia.

80. *Réalités*, November 17, 1967. National Archives of Cambodia.

81. One story was *Chnum Aun 16* (roughly *Dear of Year 16*), which is disseminated and listened to on YouTube today. This story depicts Nimul, a sixteen-year-old who lives in the outskirts of Battambang, who is tempted by the entrapments of "modern civilization," such as new cars, modern clothes, foreign films, rock and roll music, and dating wealthy young men. She gets overwhelmed by this "foreign attitude" and leaves her hometown to date a wealthy man who ends up trying to rape her. She realizes that she was being a "cheap person" instead of a traditional woman with values and marries her original boyfriend from her hometown. They have a child and he joins the military. Interspersed throughout the story is music from Sinn Sissamouth, the most famous musician from Battambang. Near the end of the story—as Nimul grapples with her own internal conflict—monk chants are powerfully collaged over Khmer rock and roll.

82. Muan and Daravuth, "A Survey of Film in Cambodia."

83. Program from the 1st Southeast Asian Biennial Film Festival and Photo Exhibition, Phnom Penh, March 29, 1997–April 5, 1997, National Archives of Cambodia.

84. "In the Service of Khmer Cinema," Som Sam Al interview and feature, *Réalités*, November 11, 1966, National Archives of Cambodia.

85. Muan and Daravuth, "A Survey of Film in Cambodia." Also in "Kon: The Cinema of Cambodia," 2010, Box 813, National Archives of Cambodia.

86. "In the Service of Khmer Cinema."

87. "In the Service of Khmer Cinema."

88. For example, Sihanouk's film *Twilight* oscillates between these two themes in visual imagery.

89. There were around thirty cinemas in the Sangkum Reastr Niyum period in Phnom Penh. From "Kon: The Cinema of Cambodia," 2010, Box 813, National Archives of Cambodia.

90. From an interview with Mao Ayuth. Lay Sovan, a businessman from Takeo, remembers that "in the countryside, when people went to watch movies at the pagoda, they always brought a mat to spread out on the ground and watched until midnight." "Kon: The Cinema of Cambodia."

91. "Kon: The Cinema of Cambodia."

92. "Kon: The Cinema of Cambodia."

93. "Kon: The Cinema of Cambodia." Also interview with Loak Chayy in Battambang, one of the painters for cinema signs.

94. Many of the soundtracks have remained when films have gone missing. "Kon: The Cinema of Cambodia."

95. Muan and Daravuth, "A Survey of Film in Cambodia."

96. Interview with Ly Bun Yim in *Golden Slumbers*, film by Davy Chou.

97. Interview with Yvon Hem in *Golden Slumbers*. See also "Kon: The Cinema of Cambodia."

98. Austin, "Gender and the Nation in Popular Cambodian Heritage Cinema."

99. "Kon: The Cinema of Cambodia."

100. Buchsbaum, "A Closer Look at Third Cinema."

101. Even Sihanouk's films were feature films made primarily for entertainment and pleasure, though they (of course) sold a particularly pleasant and unidimensional portrait of Cambodia.

102. Muan and Daravuth, "A Survey of Film in Cambodia."

103. "Films sovietiques," *Réalités*, April 8, 1966, National Archives of Cambodia.

104. *Réalités*, October 28, 1966, National Archives of Cambodia.

105. On their return they brought with them Prince Sihamoni for schooling, who is now King of Cambodia. "Cineastes checoslovaques," *Réalités*, November 25, 1966, National Archives of Cambodia.

106. "Gala de films roumains," *Réalités*, November 25, 1966, National Archives of Cambodia.

107. *Golden Slumbers* film.

108. Muan and Daravuth, "A Survey of Film in Cambodia." Also noted in the film *Golden Slumbers* and in "Kon: The Cinema of Cambodia."

109. *Réalités*, May 17, 1966, National Archives of Cambodia.

110. "Films etrangers," *Réalités*, October 13, 1967, National Archives of Cambodia.

INTERLUDE: WARTIME MEDIA

1. "Antigovernment Radio transmitter Operating in Phnom Penh," January 23, 1975, from the FBIS archive. The text reads, "An antigovernment Red Khmer radio transmitter is operating secretly somewhere inside the Cambodian capital . . . giving orders to Red Khmer units fighting on the outskirts of the capital."

2. "Corruption of Present SRV Regime Revealed," Phnom Penh Domestic Service in Cambodian, 2300 GMT, July 30, 1978. Sample text: "The present Vietnamese regime of the Le Duan-Pham Van Dong clique is not in the least revolutionary or socialist. Its true nature is corrupt, rotten, and antirevolutionary. . . . While the husbands hold the power, the wives use that power to control and own the state and people's property, disposing of it as they wish. . . . They have become utterly rotten and have systematically practiced deceit." FBIS archive.

3. Swan, "Voice of America, a Radio Heard in Secret."

4. Swan, "Voice of America, a Radio Heard in Secret."

5. This scholarship includes Becker, *When the War Was Over*; Chandler, *Brother Number One*; Hinton, *Why Did They Kill?*; and Kiernan, *The Pol Pot Regime*.

6. Beverly, *Testimonio: On the Politics of Truth*; Sarkar and Walker, "Introduction: Moving Testimonies."

7. Chhim, "Baksbat (Broken Courage)," 9.

8. Radstone, "Trauma Studies."

9. Radstone, "Trauma Studies."

10. Kleinman and Kleinman, "The Appeal of Experience."

11. Caswell, *Archiving the Unspeakable*.

12. Boyle, "Trauma, Memory, Documentary."

13. Caswell, *Archiving the Unspeakable*.

14. Gottesman, *Cambodia after the Khmer Rouge*.

CHAPTER 2

1. Vuth, "Knowledge Sharing and Learning Together."

2. For more sources on the PRK period, see Deth, "The People's Republic of Kampuchea 1979–1989"; Gottesman, *Cambodia after the Khmer Rouge*; Slocomb, *The People's Republic of Kampuchea, 1979–1989*.

3. I focus on radio and film because these were by far the most common forms of media during this period. As during the Sangkum Reastr Niyum period, there

NOTES

was some TV use in Cambodia but it was not very common, nor was it regularly used as a personal device. The Phnom Penh–based TV station did broadcast again in 1983 and Russians gave the Ministry of Information some black and white TVs. Bou Vannarith said that 90 percent of Cambodians didn't have TV or electricity in the 1980s—plus the TV played only two hours in the evening. He said that "they just had the news and it was short . . . most people did not have money for buying televisions, and they didn't have electricity." By 1990, there were also three regional television stations, in Battambang, Kampong Cham, and Sihanoukville.

4. Star and Strauss, "Layers of Silence, Arenas of Voice."

5. Jackson, "Rethinking Repair."

6. In this phrasing, I am inspired by Hecht's related concept of technopolitics, which she defines as "the hybrid forms of power embedded in technological artifacts, systems and practices" or "the strategic practice of designing or using technology to enact political goals." Hecht, *Entangled Geographies*.

7. The FBIS is an online archive of CIA-recorded and translated radio programs from Cambodia from the Cold War.

8. He also told me, "We moved to this building in 2003 for broadcasting too. When we moved here, we constructed a new studio next to this building."

9. "FBIS Monitoring Note on Phnom Penh Domestic Radio," 1979, FBIS archive.

10. "Government Leaders, SRV Experts Attend Opening of Radio Course," BK281042 Phnom Penh Domestic Service in Cambodian, 1100 GMT, September 23, 1979, FBIS archive. My emphasis.

11. Broadcasts and song programs increase in number in the first half of the 1980s. Figures from "Situation of the Implementation of the Phnom Penh Radio and Activities of the Phnom Penh Newspapers from 1979–1983" show that the number of broadcasts grew from zero to 1,590 between 1979 and 1983, and the number of song programs from 0 to 500 in the same period. *Statistics of Economics and Culture of Phnom Penh 1979–1983* (translated from Khmer). No textual contextualization provided. National Archives of Cambodia.

12. "Phnom Penh Radio to Begin English, French, Thai Services on 20 July," BK190508, July 19, 1979, translated into English and published in Daily Report, Asia & Pacific, Kampucha, FBIS archive.

13. "Phnom Penh Radio Director General Greets Foreign Listeners," BK211008, July 20, 1979, translated into English and published in Daily Report, Asia & Pacific, Kampucha, July 23, 1979, FBIS archive.

14. "Government Leaders, SRV Experts Attend Opening of Radio Course," my emphasis.

15. "Government Leaders, SRV Experts Attend Opening of Radio Course."

16. "Government Leaders, SRV Experts Attend Opening of Radio Course."

17. Phnom Penh Domestic Service in Cambodian, 1230 GMT, April 12, 1981, BK, FBIS archive.

18. "Vietnamese Technicians Help Set up Radio Station in Takhmau," BK141252, October 3, 1979, translated into English and published in Daily Report Asia & Pacific, October 26, 1979, FBIS archive.

19. "Battambang Radio Station," Phnom Penh Domestic Service in Cambodian, January 20, 1981, FBIS archive; "Provincial Radio Operation Part of Prey Veng Development," BK291241 Phnom Penh Domestic Service in Cambodian, 1200 GMT, February 26, 1980, BK, FBIS archive.

20. Figures from "Situation of the Implementation of the Phnom Penh Radio and Activities of the Phnom Penh Newspapers from 1979–1983," in *Statistics of Economics and Culture of Phnom Penh 1979–1983*, National Archives of Cambodia.

21. "Khmer Rouge Radio Transmitter in Yunnan," 1981, FBIS archive.

22. "Briefs," Bankgkok SU Anakhot in Thai, July 17–23, 1982, p. 33, BK, FBIS archive.

23. "Cambodia: Khmer Rouge Radio Transmission Cut," BK2108051496 Hong Kong AFP in English, 0447 GMT, August 21, 1996, FBIS archive.

24. "AFP: KPNLF Radio Station Begins Broadcast," BK260813 Hong Kong AFP in English, October 26, 1982, FBIS archive.

25. "Sihnaouk-Son Sann to Operate Joint Radio Station," BK140316 Bangkok Post in English, November 14, 1983, FBIS archive; "Coalition Partners to Pen Radio Station 16 Dec.," BK120130 Bangkok Post in English, December 12, 1983, p. 3, FBIS archive.

26. "Mobile Resistance Radio to Extend Range," BK180200 Bangkok *The Nation Review* in English, January 18, 1984, FBIS archive.

27. "Coalition Partners to Pen Radio Station 16 Dec.," BK120130 Bangkok Post in English, December 12, 1983, p. 3, FBIS archive; "AFP Cites CGDK New Radio Station," BK231131 Hong Kong AFP in English, 0937, January 24, 1984, FBIS archive.

28. "Mobile Resistance Radio to Extend Range," BK180200 Bangkok *The Nation Review* in English, January 18, 1984, FBIS archive.

29. "AFP Cites CGDK New Radio Station," BK231131 Hong Kong AFP in English, 0937 January 24, 1984, FBIS archive.

30. Marston, "Cambodian News Media in the UNTAC Period."

31. "Preparing for the Cambodian New Year Celebration with Responsibility and Inspiration," 1984, which was part of my "Listening from the Archives" program,

is one example. John Marston argues that "in the 1980s some quite moving songs were recorded, songs that genuinely resonated with the public's memories of the suffering during the Pol Pot period. These were broadcast by radio arts teams at a time when memories of the DK regime were still fresh and when there was still energy and excitement about reviving radio in the country. The quality of radio arts declined as the surviving pre-1975 equipment deteriorated and as artists left for greener pastures—some of them going to the camps on the border." Marston, "Cambodian News Media in the UNTAC Period."

32. See also Reddick and Taing, "After the Khmer Rouge."

33. "VOK Comments on PRK Radio's Appeal Program," BK040709 Voice of the Khmer in Cambodian, 0500 GMT, April 4, 1987, FBIS archive.

34. Shawcross, *The Quality of Mercy*.

35. See also Gottesman, *Cambodia after the Khmer Rouge*.

36. Marston also mentions in a footnote the common practice of having loudspeakers in villages during the 1980s in "Cambodian News Media in the UNTAC Period."

37. Loudspeaker networks become more robust in the first half of the 1980s. Figures from "Situation of the Implementation of the Phnom Penh Radio and Activities of the Phnom Penh Newspapers from 1979–1983," in *Statistics of Economics and Culture of Phnom Penh 1979–1983* (translated from Khmer). No textual contextualization provided. National Archives of Cambodia.

38. "SRV-Aided Radio Station," FBIS archive.

39. He and his neighbors would also listen to their personal radios together in groups at night at home. He told me that his kids weren't allowed to listen to the news, only *lkhoan niyeay* (spoken theater). They listened to the radio every weekend as a family when they played *lkhoan niyeay* or *lkhoan ayai* (spoken comedy).

40. "Hun Sen Praises Radio Achievements in Speech," BK1112130188 Phnom Penh domestic service in Cambodian, 1300 GMT, December 10, 1988, FBIS archive.

41. See "SRV Radio, TV Delegation arrives," May 31, BK010914 Phnom Penh Domestic Service in Cambodian, 0400 GMT, June 1, 1982, FBIS archive; "Radio Broadcasting Protocol Signed with SRV," BK2405154990 Phnom Penh SPK in French, 1110 GMT, May 24, 1990, FBIS archive.

42. "Trade, Radio Delegations Leave for Moscow," BK280835 Phnom Penh SPK in French, 1439 GMT, January 27, 1983, FBIS archive; "Radio Delegation to CSSR," Phnom Penh Domestic Service in Cambodian, 1100 GMT, March 8, 1984, BK, FBIS archive.

43. "CCSSR Radio Equipment," Phnom Penh SPK in French, 0429 MT, May 22, 1983, FBIS archive.

44. "Radio, Television Protocol Signed with USSR," BK30736 Phnom Penh Domestic Service in Cambodian, 1100 GMT, March 22, 1984, FBIS archive; "Radio-TV Protocol Signed with USSR 28 Jan.," BK311310 Phnom Penh SPK in French, 1153 GMT, January 31, 1987, FBIS archive; "Soviet Radio-TV Delegation Leaves 28 February," BK0403073790 Phnom Penh SPK in French, 0357 GMT, March 4, 1990, FBIS archive.

45. Some detail found in Baumgärtel, "A Profile of Mao Ayuth."

46. "Pen Sovan Reopening of Phnom Penh Movie Theater," BK031226 SPK in French, 0418 GMT, January 6, 1980, BK, FBIS archive.

47. "SRV-Aideed Radio Station," Phnom Pnh SPK in French, 0404 GMT, February 11, 1982, FBIS archive.

48. "Activities of the Department of Cultural Propaganda," 1979–1983, in *Statistics of Economics and Culture of Phnom Penh 1979–1983*, National Archives of Cambodia.

49. "Kon: The Cinema of Cambodia."

50. Held at the Bophana Center.

51. "The Season of the Palm Flowers" is just one of more than a dozen short films made by Ieu Pannakar and Mao Ayuth during this time held at the Bophana Center. *Kampuchea Rsa Laengvinh* (Rebirth of Kampuchea) (1985), by Ieu Pannakar and Nget Samorn, is a 36-minute documentary about the post–Pol Pot era. The film opens with images of the university and school structures, which are functioning again; the film offers thanks to the help of Vietnamese and Russians. The documentary also shows the restoration of Angkor Wat and the travel of Cambodian students to Moscow in an exchange program. Juxtaposed to these images of recovery are pictures of the Pol Pot–Ieng Sary regime. Films are held at the Bophana Center.

52. According to a brochure for the first Southeast Asian biennial film festival and photo exhibition in 1997, the Cambodian National Assembly allowed filmmakers to resume independent activities in 1988. National Archives of Cambodia.

53. For more, see Baumgärtel, "A Profile of Mao Ayuth."

54. This film is held at the Bophana Center.

55. Now known in English as the Bat Cave, a popular tourist destination for seeing bats at sunset.

56. Translated from Khmer with Nehru Ski.

57. These collectors include Preah Sorya, the subject of the fourth chapter. Other interview participants have told me that Ka Toy was involved in exporting original film reels to Long Beach with Sam Sovandeth, who would then transfer them to cassettes and resell the films. This story is controversial because Sam Sovandeth was arrested in 2010. Sovandeth had a large number of media businesses spanning Long

Beach and Cambodia, and he was an official for a new radio and television network called Southeast Asia Television and Radio in conjunction with a government official named Kao Kim Hourn. He is also the owner of one of the remaining cinemas in Phnom Penh called Lux Cinema. In May 2010 Sovandeth was convicted for stealing $4.7 million from the television station, and (according to reports) he was in prison in 2018. The unavailability of Sam Sovandeth to talk has come up as a difficult issue for many contemporary classical Khmer film enthusiasts.

58. "Kon: The Cinema of Cambodia."

59. Gottesman, *Cambodia after the Khmer Rouge*.

60. In April 1986, the Council of Ministers issued the new Subdecree 9 and distinguished between criminal offenses that involved "political security" and unauthorized showings of videos "for commercial purposes." The prosecutor on the case tried the two Yens under this new Subdecree and punished them with a fine and confiscation of the videos. The police disagreed, arguing that the "two Yens" had compromised state security, and kept the Yens in custody. The prosecutor rejected their opinion and said that their file had already been sent to the Ministry of Justice and requested the release of the two Yens. The police appealed to the Phnom Penh municipal authorities and the city's Party Committee. The authorities asked that the police and prosecutor resolve the issue on their own. Vietnamese advisors stepped into mediate, but various advisors also disagreed about how to proceed. The case was in a stalemate until August 1986, when it magnified into a "fundamental question of institutional authority" and played out with complicated politics. Heng Samrin, on the Politburo, said, "If we rely on the law, it is not necessary to put these people in jail." Say Pnouthang added, "Only videocassettes that broadcast Sihanouk or Khieu Samphan [Khmer Rouge leadership] are included in cases of political security." The Politburo, however, watched the pornographic videos together, and Heng Samrin revised his opinion. "The Politburo has now determined that the case serves the psychological warfare of the enemy." There are no further meetings on record, and the case seems to have faded away. In Gottesman, *Cambodia after the Khmer Rouge*.

61. Nelson, *Reckoning*.

INTERLUDE: PEACE TALKS

1. "Safeguarding Peace: Cambodia's Constitutional Challenge."

2. Heder and Ledgerwood, "Introduction."

3. "Safeguarding Peace: Cambodia's Constitutional Challenge."

4. Heder and Ledgerwood, "Introduction."

5. Heder and Ledgerwood, "Introduction."

6. "Safeguarding Peace: Cambodia's Constitutional Challenge."

7. Heder and Ledgerwood, "Introduction."

8. Heder and Ledgerwood, "Introduction."

9. Heder and Ledgerwood, "Introduction."

10. "Safeguarding Peace: Cambodia's Constitutional Challenge."

11. "Safeguarding Peace: Cambodia's Constitutional Challenge."

CHAPTER 3

1. While in Wisconsin, I worked closely with Lawrence Ashmun, the head librarian for the Southeast Asian studies collection. He was able to walk me through the catalogue of the 333 radio cassettes, 350 reels, 1,417 digital audio tapes, 12 audio packs, and 2,464 paper documents (including transcripts from the radio and internal UNTAC letters, among other documents) held in their library from the UNTAC era, uniquely held by the University of Wisconsin Library (the original material produced by UNTAC, as received straight from them). Most important to me were the letters, histories of Radio UNTAC, and radio program schedules in the miscellaneous paper files in the collection. I was also able to digitize radio programs (including one called "The History of Radio UNTAC") with the help of Dorothea Salo at RADD (Recovering Analogue and Digital Data) in the Wisconsin Information and Library School.

2. Marston, "Cambodian News Media in the UNTAC Period."

3. Marston, "Cambodian News Media in the UNTAC Period."

4. Although newspapers are not a focus of this project, small, independent, Khmer businesses or independent institutions funded by foreign capital started Cambodian newsletters from Phnom Penh (not just from the Thai border) during the UNTAC period. In 1992 two English-language newspapers—the *Cambodia Times* (financed by a Malaysian business) and the *Phnom Penh Post* (financed by an American business)—began. *Cambodia Times* also had a Khmer-language equivalent. The *Cambodia Times* was "slickly edited" and sometimes partial to the CPP. The third English newspaper—the *Cambodia Daily*—began in 1993. Though the *Cambodia Times* closed in 1997, the *Cambodia Daily* remained in circulation until 2017, and the *Phnom Penh Post* remains the premier English-language newspaper in Cambodia, though it was bought by a Malaysian company with ties to the CPP in 2018. In early 1993, other small, independent, Khmer-language newspapers started appearing. Nonstate media represented a departure from the PRK and harkened back ideas about the old regime and the time before communist media. For example, the *Koh Santepheap* (Island of Peace), a newspaper that ran in the Sihnaouk and Lon Nol eras, was restarted in

NOTES 213

January 1993 by a reporter for the journal from the 1970s. Its first editor was killed by the Khmer Rouge in 1975.

TV also opened more channels; aside from the one Cambodian national station (TV Kampuchea, which started broadcasting again in very limited hours in 1986), in early 1993, FUNCINPEC opened a second TV station. Shortly afterward, IBC TV—a Thai corporate TV station with income from advertising—began programming. It had a broadcasting radius of 100 kilometers around Phnom Penh and a relay station in Kampong Cham. It played movies from the United States and game shows from Hong Kong, dubbed in Khmer. It became popular very quickly; Thai-speaking Cambodians who had returned from the border areas made up most of the staff. For more, see Heder and Ledgerwood, "Introduction"; Marston, "Cambodian News Media in the UNTAC Period."

5. Heder and Ledgerwood, "Introduction."

6. The "Voice of the Khmer People" was forced to stop broadcasting in 1992 because Thailand and other countries financially supporting the station were "under pressure to maintain neutrality with respect to the elections." Heder and Ledgerwood, "Introduction."

7. Heder and Ledgerwood, "Introduction."

8. Heder and Ledgerwood, "Introduction."

9. Steven Pak is one example of a Cambodian refugee who spent 10 years in California and returned to Cambodia to work on Radio UNTAC. Zhou, *Radio UNTAC of Cambodia*.

10. Jeffrey Heyman, "The History of Radio UNTAC," translated by Stephen Pak, September 22, 1993, UNTAC Collection.

11. Heyman, "The History of Radio UNTAC."

12. Discussed in chapter 1.

13. Zhou, *Radio UNTAC of Cambodia*.

14. Heyman, "The History of Radio UNTAC."

15. Zhou, *Radio UNTAC of Cambodia*.

16. Heyman, "The History of Radio UNTAC."

17. Lyno Vuth, a prominent contemporary artist, has told me he finds the appropriation of this culturally significant music disrespectful.

18. "UNTAC Radio Winding Up," *Cambodia Times*, September 27–October 3, 1993, National Archives of Cambodia.

19. "Good Morning Cambodia," *Cambodia Times*, June 7–13, 1993, National Archives of Cambodia.

20. "UNTAC Radio Winding Up."

21. Heyman, "The History of Radio UNTAC."

22. Zhou, *Radio UNTAC of Cambodia*.

23. Zhou, *Radio UNTAC of Cambodia*.

24. "Grenades in Exchange for Radios," *Cambodia Times*, June 7–13, 1993, National Archives of Cambodia.

25. Zhou, *Radio UNTAC of Cambodia*.

26. There was no distribution in Phnom Penh.

27. Zhou, *Radio UNTAC of Cambodia*.

28. Zhou, *Radio UNTAC of Cambodia*.

29. "Grenades in Exchange for Radios."

30. Heyman, "The History of Radio UNTAC."

31. "UNTAC Radio Winding Up."

32. Heyman, "The History of Radio UNTAC"; Zhou, *Radio UNTAC of Cambodia*; "Good Morning Cambodia."

33. Heyman, "The History of Radio UNTAC."

34. Marston, "Cambodian News Media in the UNTAC Period."

35. Zhou, *Radio UNTAC of Cambodia*.

36. "UNTAC Radio Winding Up."

37. Zhou, *Radio UNTAC of Cambodia*.

38. He continued, "On the first day of the elections, UNTAC radio reported on UNTAC's fines against Prince Norodom Chakkrapong and Khim Bo. Later, it broadcast an interview with a Buddhist monk, in which SOC provincial officials were accused of barring people from offering him food. Moreover, the UNTAC radio also belittled Prince Norodom Chakkrapong and Khim Bo when they paid their fines. During the elections, we also noted that UNTAC radio played FUNCIPNPEC's political songs." Khieu Kanhari "underlined that this occurrence is a lesson to other countries, which like Cambodia accept the UN presence, to pay careful attention to the news service because it can greatly influence an election."

39. "UNTAC Radio Winding Up."

40. The Cambodian ruling party still makes statements about their resentment of the role of UNTAC and the legacy of the Paris Peace Accords. Hun Sen publicly said in March 2018 that Cambodians were responsible for the national reconstruction

and that the foreign histories of the Paris Peace Accords are largely flawed. See Sokhean, "Hun Sen: Peace Brought by Khmers, Not 'Foreign Hands.'"

41. Heder and Ledgerwood, "Introduction."

42. Marston, "Cambodian News Media in the UNTAC Period." UNTAC (specifically, the Information/Education Division) had tried to start a media association, but this group did not gain traction and authority. International journalists, entrenched during the UNTAC period, were able to successfully establish it independently by the end of 1993. It had limited power but had a major success when, in late 1993, the SOC announced the old SOC law on the press was valid. After the association protested, the state negotiated. In March 1994, the CPP announced an FM radio station and a new TV station. Many smaller periodicals closed at the elections and some never reopened. In this same month FUNCINPEC Radio and TV became private. FUNCINPEC radio became the most popular after the closing of Radio UNTAC.

43. Marston, "Cambodian News Media in the UNTAC Period."

44. "Rapid Economic Growth Brings Investors," *Cambodia Times*, September 27–October 3, 1993, National Archives of Cambodia.

45. "Rapid Economic Growth Brings Investors."

46. In Battambang, within a few years after UNTAC, Hab B became a nightclub and Hap Chouen became a restaurant.

47. After 1992, for example, it became more difficult for Ka Toy to do his job. With the projector change, it became harder to find films to play on the film projectors that he had. As VCRs became more available, fewer people came to the cinema. Ka Toy thinks part of the reason people stopped coming to the cinema was because the quality of the films also went down. The LCD projectors made a lot of noise and the color was less clear.

48. 2002 Internet Cambodia Case Study, International Telecommunication Union, National Archives of Cambodia.

49. In 1993, Cambodia was the first country in the world where mobile telephone subscribers surpassed fixed ones. 2002 Internet Cambodia Case Study.

50. 2002 Internet Cambodia Case Study.

51. 2002 Internet Cambodia Case Study.

52. 2002 Internet Cambodia Case Study.

53. 2002 Internet Cambodia Case Study.

54. 2002 Internet Cambodia Case Study

55. 2002 Internet Cambodia Case Study.

56. From interviews with Norbert Klein and Moa Chakrya.

57. Jack, Sovannaroth, and Dell, "Privacy Is Not a Concept."

58. For instance, the lack of appropriate Burmese language content moderation on Facebook in Myanmar arguably exacerbated the genocide against the Rohingya people.

CHAPTER 4

1. One of the most famous was called *A Prison without Walls*, which was about, Chayy explains, "when the Khmer Rouge took power, the life of people during that time, how they treated the people, the cruelty. The play went until liberation day."

2. Simmel, "Two Essays (The Handle, The Ruin)."

3. Stoler, "Imperial Debris."

4. Derrida, *Specters of Marx*.

5. Derrida, in *Specters of Marx*, gives the example of the absence of recognition of the death of workers in a mining accident.

6. Gordon, *Ghostly Matters*.

7. Williams and Orrom, *Preface to Film*; Berlant, "Cruel Optimism."

8. Kittler, *Gramophone, Film, Typewriter*.

9. This is a theme I will discuss in greater detail in the next chapter.

10. See chapter 1 for more on this theme.

11. Sereypagna, "New Khmer Architecture."

12. United Nations Development Program Cambodia Profile 2019, https://hdr.undp.org/en/countries/profiles/KHM.

13. Beech, "Cambodia Re-Elects Its Leader, a Result Predetermined by One."

14. The *Phnom Penh Post* staff, "Sokha Arrested for Treason, Is Accused of Colluding with US to Topple the Government."

15. Mech and Baliga, "Death of Democracy."

16. For more on Hun Sen's approach to ruling, see Norén-Nilsson, *Cambodia's Second Kingdom*.

17. Reporters without Borders, "Cambodia: The Independent Press in Ruins."

18. Reuters, "Cambodia Daily Shuts with 'Dictatorship' Parting Shot at Prime Minister Hun Sen"; Reporters without Borders, "Cambodia: The Independent Press in Ruins."

19. Vichheika, "RFA Journalists Accused of Treason."

20. Cambodian Center for Independent Media, "Challenges for Independent Media."

21. Radio Free Asia staff, "Cambodia to Monitor, Control Online News."

22. Reuters, "Cambodia Blocks Some Independent News Media Sites."

23. Vong and Hok argue that everyday youth Facebooking in Cambodia had a strong influence on Cambodian politics and government action in the 2013 elections. Though everyday youth Facebooking still plays an important role in Cambodian politics, Vong and Hok's paper was written in the wake of the 2013 election. When I conducted this ethnography, the political realities and Facebook freedoms had changed. Vong and Hok, "Facebooking."

24. Vong and Hok, "Facebooking."

25. Hughes and Eng, "Facebook, Contestation and Poor People's Politics."

26. Beban, Schoenberger, and Lamb, "Pockets of Liberal Media."

27. Beban, Schoenberger, and Lamb, "Pockets of Liberal Media."

28. Gordillo, *Rubble*, 28.

29. Gordillo, *Rubble*.

30. Benjamin, *The Origin of German Tragic Drama*.

31. Gordon, *Ghostly Matters*.

32. See also Johnson, *Ghosts of the New City*.

33. Choulean, *People and Earth*.

34. Schwenkel, "Haunted Infrastructure."

35. The team told me they could not find too much information at institutional archives such as the National Archives. One of their most important sources was Amazing Cambodia, a Facebook page run by Srin Sokmean, which I discuss in chapter 6. I once went to the National Museum to look at the Ingrid Muan papers with them.

36. The tool is called Bosch Laser Measurement.

37. AutoCAD is free for three years for students.

38. Khmer Architecture tours is a private tour company geared toward expats, tourists, and researchers. Limkokwing is a private and expensive Malaysian arts and architecture school in Phnom Penh.

39. In 2019, she married and was hired on an urban heritage project in Phnom Penh.

40. Java Cafe had three locations across Phnom Penh in 2017–2018. Sa Sa Bassac was described by Meta as the only "proper gallery" in Phnom Penh; it closed in 2018.

CHAPTER 5

1. Hong Saovandy, a government official at the Ministry of Transportation, shares a common opinion: "I prefer the movies that were made at that time to those produced today. I feel that the movies from the 1960s look and feel natural, and the actors really made an effort to impersonate what they were playing." From "Kon: The Cinema of Cambodia."

2. "Kon: The Cinema of Cambodia."

3. One person who has been a forerunner in the preservation of Cambodian heritage film is Davy Chou, whom I don't explicitly discuss but has been an inspiration for Preah Sorya and other young Cambodian collectors. He is also one of the founders of the Anti-Archive film production company. Davy Chou made *Golden Slumbers* in 2010, the most comprehensive study of the 1960s film period to date and which I reference throughout this chapter.

4. Translation by author.

5. In Khmer he said *kduk kdoul*, translated as a very strong feeling similar to the English word "shocked."

6. Dubois et al., "Household Survey of Psychiatric Morbidity in Cambodia"; De Jong, Komproe, and Van Ommeren, "Common Mental Disorders in Postconflict Settings."

7. Shulevitz, "The Science of Suffering."

8. Um, *From the Land of Shadows*, 7. She works here from Derrida, *Specters of Marx*.

9. Chhim, "Baksbat (Broken Courage)."

10. Chhun, "Walking with the Ghost."

11. Chhun, "Walking with the Ghost," 26.

12. Thompson, "Forgetting to Remember, Again."

13. Uk, *Salvage*.

14. Hinton, *The Justice Façade*.

15. Schlund-Vials, *War, Genocide, and Justice*.

16. Hinton, *The Justice Façade*, 85–86.

17. For more on the embodied aspects of trauma, see Van der Kolk, *The Body Keeps the Score*.

18. Eisenbruch, "From Post-Traumatic Stress Disorder to Cultural Bereavement."

19. Guillou explains that her research is a necessary corrective to theories of trauma developed in part by groups like the Extraordinary Chambers of the Courts of Cambodia. She says, "The accounts given in many areas of the Tribunal have tended to impose an overall pattern of what social suffering should be and how it should be expressed and relieved, by using the idioms of 'trauma' and 'post-traumatic stress disorder' as their only models. The ideology underlying the overuse of the psychiatric scheme helps to reinforce the marginalization of small societies such as Cambodia in the globalized world by producing a particular image of them: not only did the Cambodians slaughter each other during an 'auto-genocide,' not only did they suffer from mental illness caused by trauma, but they also remained unconcerned and passive after the genocide. This ideological pattern in turn aims to make the proliferation of so-called humanitarian organizations in Cambodia acceptable by spreading the idea that Cambodians cannot help themselves or be treated as responsible citizens." From Guillou, "An Alternative Memory of the Khmer Rouge Genocide."

20. Choulean, *People and Earth*.

21. Uk, *Salvage*.

22. Schlund-Vials and Ly recognize some (limited) benefits of applying Western trauma theory to describe the Cambodian experience of painful experience and for textual analysis of Cambodian commemorative art production. In the late 1990s, the medical and popular use of the trauma concept was taken up by literary scholars and then branched out to become a subdiscipline within the humanities called trauma studies, spearheaded by Cathy Caruth. Caruth's understanding of the repetitive and uncertain nature of traumatic memory informed my theorization of disintegration noise, but is less relevant to the Cambodian experience of surviving the Khmer Rouge, so I decided to eliminate it from the body of this text. For more, see Caruth, *Unclaimed Experience*. To see this type of analysis applied to Southeast Asian experiences of trauma, see Boyle, "Trauma, Memory, Documentary."

23. Schlund-Vials, *War, Genocide, and Justice*.

24. Ly, *Traces of Trauma*.

25. Elsaesser, "Freud and the Technical Media."

26. Benjamin, *The Work of Art in the Age of Mechanical Reproduction*.

27. McLuhan, *Understanding Media*; Kittler, *Gramophone, Film, Typewriter*.

28. Kittler, *Gramophone, Film, Typewriter*, 13.

29. Parikka, "Mapping Noise."

30. Strassler's concept of "refraction" also gives insight into the ways that all mechanical reproduction can both mimic and distort. Strassler, *Refracted Visions*.

31. Parikka, "Mapping Noise." See also Nunes, "Error, Noise, and Potential."

32. Kelly, *Cracked Media*.

33. Krapp, *Noise Channels*.

34. Menkman, "Glitch Studies Manifesto."

35. Steyerl, "In Defense of the Poor Image"; Larkin, *Signal and Noise*.

36. This condition harkens back to Guillou's concept of the "switch on/switch off" of remembering and forgetting. Films flick memory back on for a discrete time, and can be turned off again.

37. Here I am inspired in part by Hirsch, who argues that historical (pre-trauma) photographs allow those suffering from events of collective trauma to move on. Photographs can "absorb the shock, filter and diffuse the impact of trauma, diminish harm." Hirsch, "The Generation of Postmemory."

38. Kong Sam Oeun was an extremely productive and popular actor, who starred in 140 films. His first movie was *Koth Muoy Py Neak* (One Coffin Two People), and he performed more than 200 songs. "Kon: The Cinema of Cambodia."

39. Boeng Keng Lake was once in Phnom Penh but a company purchased it for redevelopment in 2011 and filled in in.

40. As noted in the previous chapter and as I will discuss again in the following chapter, this commemoration practice is not neutral. The period they romanticize was one of deep inequality. What they choose to focus on (the 1960s and early 1970s film culture) was also deeply embedded in imperial and elite politics. Further, as I discuss in greater depth in the next chapter, the group focuses on memories that refer to the trauma of the wars without directly referencing them. The group focuses not on memories from the traumatic event (the Khmer Rouge era and wartime) but instead on the positive cultural outputs from the period before that. The students are able to commemorate lost artists without focusing on violence; they thus decenter what they often perceive to be a simplification of their national history through an international focus on the Khmer Rouge period.

41. See the previous chapter for a longer discussion on this impulse through the case of Roung Kon.

42. Crownshaw, Kilby, and Rowland, *The Future of Memory*.

NOTES

43. Hedges, *World Cinema and Cultural Memory*.

44. Harvey, "Memory, That Powerful Political Force."

CHAPTER 6

1. See https://www.facebook.com/amazingcambo.

2. There were 96,648 as of August 8, 2022.

3. Chun, "The Enduring Ephemeral"; Chun, *Programmed Visions*.

4. Working from the definition of the "imperial archive" as defined by Richards in *Imperial Archive*.

5. Though many Cambodians increasingly use Khmer script on Facebook, it is common to see a mix of English, Khmer script, and Romanized Khmer on many Cambodian Facebook pages, since often roman scripts are easier to type with and are more globally accessible. Cambodia is also famously known as a large market for audio messages due to difficulty typing in Khmer. See Elliott and Phorn, "Fifty Percent of Facebook Messenger's Total Voice Traffic."

6. Though he does sometimes find some information at the National Archives in Phnom Penh.

7. This magazine ran from 1927 through 1975 and recently restarted.

8. Chun, "The Enduring Ephemeral"; Chun, *Programmed Visions*.

9. Hooper, "Green Computing."

10. She says, "A memory must be held to keep it from moving or fading." Chun, "The Enduring Ephemeral," 164.

11. Weltevrede, Helmond, and Gerlitz, "The Politics of Real-Time."

12. As Derrida has famously claimed, "there is no political power without control of the archive," *Archive Fever*. For another take on archiving in Cambodia, about the photographs of Tuol Sleng victims, see Caswell, *Archiving the Unspeakable*. Caswell has also written with Ricardo Punzalan about the function of the institutional archive for social justice. See Punzalan and Caswell, "Critical Directions."

13. Richards, *Imperial Archive*, 3.

14. This is what Richards calls "entropic disorganization."

15. Stoler, *Along the Archival Grain*.

16. Doreen Lee describes the grassroots archives created by the activist youths of late 1990s Indonesia who were the forefront of the Rerformasi, which toppled the

Suharto government on May 21, 1998. These youth sought to document each action and protest since they knew their perspective—their knowledge—was marginalized. Lee, *Activist Archives*.

17. Irani et al., "Postcolonial Computing." See also Irani, *Chasing Innovation*.

18. In the past, ICTD practitioners rarely did user research before spending a significant amount of money on building a new product, which then often went unused by expected beneficiaries. However, the field is largely changing and becoming more attuned to questions of politics and participation, and moving away from techno-deterministic approaches.

19. For one emblematic example, Morgan Ames criticizes One Laptop per Child in part for its inefficiency in *The Charisma Machine*.

20. For more on challenges on "global" and "local" design cultures, see Chan, *Networking Peripheries*, and Avle and Lindtner, "Design(ing) 'Here' and 'There.'"

21. Ames, *The Charisma Machine*.

22. Wyche et al., "If God Gives Me the Chance I Will Design My Own Phone."

23. Ahmed, Mim, and Jackson, "Residual Mobilities."

24. Jack, Chen, and Jackson, "Infrastructure as Creative Action."

25. See also Punzalan and Caswell, "Critical Directions."

26. The mission of Facebook to "bring the world closer together" has been co-opted in Cambodia by the increasingly authoritarian state.

27. Jack, Sovannaroth, and Dell, "Privacy Is Not a Concept."

28. Jack, Sovannaroth, and Dell, "Privacy Is Not a Concept."

29. Lackaff and Moner, "Local Languages, Global Networks."

30. Quigley et al., "Issues and Techniques in Translating Scientific Terms."

31. For more on the challenges of digitizing minority languages in Unicode, see Zaugg, "Digitizing Ethiopic."

32. Haraway, "Situated Knowledges."

33. Gillespie, "The Politics of 'Platforms'"; Gehl, "The Archive and the Processor."

34. Gillespie, "The Politics of 'Platforms.'"

35. Couldry and Mejias, "Data Colonialism."

36. We visited two more classrooms and about the same ratio applies; only three students of the forty or fifty students in class had smartphones with them.

CONCLUSION

1. Starr and Strauss, "Layers of Silence, Arenas of Voice."

2. Tufekci, *Twitter and Tear Gas*.

3. For more on affect and politics in Cambodia, see Beban, *Unwritten Rule*.

4. Jackson, "Rethinking Repair."

5. See Sinpeng and Tapsell, *From Grassroots Activism to Disinformation*, for a regional view.

6. Chandran, "Cambodia's Internet Gateway."

7. Some scholars—not without criticism—have termed this transnational political economy of platforms "data colonialism." See Couldry and Mejias, "Data Colonialism."

8. Huang, *Capitalism with Chinese Characteristics*.

9. I have encouraged, in earlier work, proper translation of user interfaces. See, for example, Jack, Sovannaroth, and Dell, "Privacy Is Not a Concept."

10. Ong and Collier, *Global Assemblages*.

11. Irani et al., "Postcolonial Computing"; De Laet and Mol, "The Zimbabwe Bush Pump."

12. Jackson, "Rethinking Repair."

BIBLIOGRAPHY

MANUSCRIPT COLLECTIONS

Bophana Center Audio-Visual Archive, Phnom Penh, Cambodia

The Center for Khmer Studies Library, Collection of *Réalités* magazines, Siem Reap, Cambodia

Foreign Broadcast Information Service (FBIS) Daily Reports, the United States' principal record of political and historical open source intelligence, consisting of Daily Reports published from September 4, 1941, through 1996, produced in cooperation with Dartmouth College Library, the Library of Congress, Tufts University, and Yale University (available online through an academic library subscription)

Ingrid Muan Papers, National Museum of Cambodia, Phnom Penh, Cambodia, Boxes 28–35

National Archives of Cambodia, Phnom Penh, Cambodia

National Radio of Cambodia, Phnom Penh, Cambodia

Radio UNTAC materials from Cambodia, 1992–1993, University of Wisconsin Library

BOOKS AND ACADEMIC ARTICLES

Ahmed, Syed Ishtiaque, Nusrat Jahan Mim, and Steven J. Jackson. "Residual Mobilities: Infrastructural Displacement and Post-Colonial Computing in Bangladesh." In *CHI 2015: Proceedings of the 33rd Annual CHI Conference on Human Factors in Computing Systems*, 427–456. New York: Association for Computing Machinery, 2015.

Aitken, Ian. "British Governmental Institutions, the Regional Information Office in Singapore and the Use of the Official Film in Malaya and Singapore, 1948–1961." *Historical Journal of Film, Radio and Television* 35, no. 1 (2015): 27–52.

Ames, Morgan G. *The Charisma Machine: The Life, Death, and Legacy of One Laptop per Child*. Cambridge, MA: MIT Press, 2019.

Anand, Nikhil. "Pressure: The Politechnics of Water Supply in Mumbai." *Cultural Anthropology* 26, no. 4 (2011): 542–564.

Austin, Jessica. "Gender and the Nation in Popular Cambodian Heritage Cinema." Master's thesis, University of Hawai'i Mānoa, 2014.

Avle, Seyram, and Silvia Lindtner. "Design(ing) 'Here' and 'There': Tech Entrepreneurs, Global Markets, and Reflexivity in Design Processes." In *CHI 2016: Proceedings of the 2016 CHI Conference on Human Factors in Computing Systems*, 2233–2245. New York: Association for Computing Machinery, 2016.

Baumgärtel, Tilman. "A Profile of Mao Ayuth." *Southeast Asian Cinema* (blog), February 2, 2011. https://southeastasiancinema.wordpress.com/2011/02/02/a-profile-of-mao-ayuth/.

Beban, Alice. *Unwritten Rule: State-Making through Land Reform in Cambodia*. Ithaca, NY: Cornell University Press, 2021.

Beban, Alice, Laura Schoenberger, and Vanessa Lamb. "Pockets of Liberal Media in Authoritarian Regimes: What the Crackdown on Emancipatory Spaces Means for Rural Social Movements in Cambodia." *Journal of Peasant Studies* 47, no. 1 (2020): 95–115.

Becker, Elizabeth. *When the War Was Over: Cambodia and the Khmer Rouge Revolution*. New York: Public Affairs, 1998.

Beech, Hannah. "Cambodia Re-Elects Its Leader, a Result Predetermined by One." *New York Times*, July 29, 2018. https://www.nytimes.com/2018/07/29/world/asia/cambodia-election-hun-sen.html.

Benjamin, Walter. *The Origin of German Tragic Drama*. Translated by John Osborne. London: Verso, 2009.

Benjamin, Walter. *The Work of Art in the Age of Mechanical Reproduction*. Translated by J. A. Underwood. London: Penguin Books, 2008.

Berlant, Lauren. "Cruel Optimism: On Marx, Loss and the Senses." *New Formations* 63 (2007): 33–51.

Beverly, John. *Testimonio: On the Politics of Truth*. Minneapolis: University of Minnesota Press, 2004.

Bong, Chansambath. "Cambodia's Disastrous Dependence on China: A History Lesson." *The Diplomat*, December 3, 2019. https://thediplomat.com/2019/12/cambodias-disastrous-dependence-on-china-a-history-lesson/.

Boyle, Deirdre. "Trauma, Memory, Documentary: Re-Enactment in Two Films by Rithy Panh (Cambodia) and Garin Nugroho (Indonesia)." In *Documentary Testimonies: Global Archives of Suffering*, ed. Bhaskar Sarkar and Janet Walker, 155–172. Abingdon: Routledge, 2010.

Buchsbaum, Jonathan. "A Closer Look at Third Cinema." *Historical Journal of Film, Radio and Television* 21, no. 2 (2001): 153–166.

Burrell, Jenna. "The Field Site as a Network: A Strategy for Locating Ethnographic Research." *Field Methods* 21, no. 2 (2009): 181–199.

Byler, Darren. *Terror Capitalism: Uyghur Dispossession and Masculinity in a Chinese City*. Durham, NC: Duke University Press, 2021.

Cambodian Center for Independent Media. "Challenges for Independent Media." 2017. https://www.ccimcambodia.org/wp-content/uploads/2018/02/Challenges-for-Independent-Media-2017-English.pdf.

Caruth, Cathy. *Unclaimed Experience: Trauma, Memory, and History*. Baltimore, MD: Johns Hopkins University Press, 1996.

Caswell, Michelle. *Archiving the Unspeakable: Silence, Memory, and the Photographic Record in Cambodia*. Madison: University of Wisconsin Press, 2014.

Chan, Anita. *Networking Peripheries: Technological Futures and the Myth of Digital Universalism*. Cambridge, MA: MIT Press, 2013.

Chandler, David P. *Brother Number One: A Political Biography of Pol Pot*. London: Routledge, 2018.

Chandler, David P. *The Tragedy of Cambodian History: Politics, War, and Revolution since 1945*. New Haven, CT: Yale University Press, 1991.

Chandran, Rina. "Cambodia's Internet Gateway Raises Fears of China-Style Surveillance." Thomas Reuters Foundation News, February 16, 2022. https://news.trust.org/item/20220216124054-u6xyw.

Chhim, Sotheara. "Baksbat (Broken Courage): The Development and Validation of the Inventory to Measure Baksbat, a Cambodian Trauma-Based Cultural Syndrome of Distress." *Culture, Medicine, and Psychiatry* 36, no. 4 (2012): 640–659.

Chhun, Lina. "Walking with the Ghost: Affective Archives in the Afterlife of the Cambodian Holocaust." *Frontiers: A Journal of Women Studies* 40, no. 3 (2019): 24–62.

Choulean, Ang. *People and Earth*. Phnom Penh: Reyum Gallery, 2000. Translated by the author from the Khmer language.

Chun, Wendy Hui Kyong. "The Enduring Ephemeral, or the Future Is a Memory." *Critical Inquiry* 35, no. 1 (2008): 148–171.

Chun, Wendy Hui Kyong. *Programmed Visions: Software and Memory*. Cambridge, MA: MIT Press, 2011.

Coombes, Annie E. *History after Apartheid: Visual Culture and Public Memory in a Democratic South Africa*. Durham, NC: Duke University Press, 2003.

Couldry, Nick, and Ulises A. Mejias. "Data Colonialism: Rethinking Big Data's Relation to the Contemporary Subject." *Television & New Media* 20, no. 4 (2019): 336–349.

Crane, Brent. "Red-Handed: Pinning the Blame for Dap Chhuon on the CIA." *Phnom Penh Post*, June 3, 2016. https://www.phnompenhpost.com/post-weekend/red-handed-pinning-blame-dap-chhuon-cia.

Crownshaw, Richard, Jane Kilby, and Antony Rowland, eds. *The Future of Memory*. Oxford: Berghahn Books, 2010.

Daravuth, Ly, and Ingrid Muan. *Cultures of Independence: An Introduction to Cambodian Arts and Culture in the 1950's and 1960's*. Phnom Penh: Reyum Gallery, 2001.

De Jong, Joop T. V. M., Ivan H. Komproe, and Mark Van Ommeren. "Common Mental Disorders in Postconflict Settings." *The Lancet* 361, no. 9375 (2003): 2128–2130.

De Laet, Marianne, and Annemarie Mol. "The Zimbabwe Bush Pump: Mechanics of a Fluid Technology." *Social Studies of Science* 30, no. 2 (2000): 225–263.

Denzin, Norman K., and Yvonna S. Lincoln. *Strategies of Qualitative Inquiry*. Thousand Oaks, CA: Sage Publications, 1994. See esp. "Chapter 1: Introduction" and "Chapter 2: The Dance of Qualitative Research Design," 1–55.

Derrida, Jacques. *Archive Fever: A Freudian Impression*. Chicago: University of Chicago Press, 1996.

Derrida, Jacques. *Specters of Marx: The State of the Debt, the Work of Mourning and the New International*. New York: Routledge, 1994.

Deth, Sok Udom. "The People's Republic of Kampuchea 1979–1989: A Draconian Savior?" Dissertation, Zaman University, 2009. Available at SSRN 2461095.

deWald, Erich. "Taking to the Waves: Vietnamese Society around the Radio in the 1930s." *Modern Asian Studies* 46, no. 1 (2012): 143–165.

Dourish, Paul. *The Stuff of Bits: An Essay on the Materialities of Information*. Cambridge, MA: MIT Press, 2017.

Dubois, Vincent, René Tonglet, Philippe Hoyois, Ka Sunbaunat, Jean-Paul Roussaux, and Edvard Hauff. "Household Survey of Psychiatric Morbidity in Cambodia." *International Journal of Social Psychiatry* 50, no. 2 (2004): 174–185.

Ebihara, May Mayko. *Svay: A Khmer Village in Cambodia*. Ithaca, NY: Cornell University Press, 2018.

Eco, Umberto. *How to Write a Thesis*. Cambridge, MA: MIT University Press, 2015. See esp. "Chapter 3, Conducting Research," 45–106.

Edwards, Penny. *Cambodge: The Cultivation of a Nation, 1860–1945*. Honolulu: University of Hawai'i Press, 2007.

Eisenbruch, Maurice. "From Post-Traumatic Stress Disorder to Cultural Bereavement: Diagnosis of Southeast Asian Refugees." *Social Science & Medicine* 33, no. 6 (1991): 673–680.

Elliott, Vittoria, and Bopha Phorn. "Fifty Percent of Facebook Messenger's Total Voice Traffic Comes from Cambodia. Here's Why." *The Rest of the World*, November 12, 2021. https://restofworld.org/2021/facebook-didnt-know-why-half-of-messengers-voice-traffic-comes-from-cambodia-heres-why/.

Elsaesser, T. P. "Freud and the Technical Media: The Enduring Magic of the Wunderblock." In *Media Archaeology: Approaches, Applications, and Implications*, ed. Jussi Parikka and Erkki Huhtamo, 95–116. Berkeley: University of California Press, 2011.

Elwood, Sarah. "Digital Geographies, Feminist Relationality, Black and Queer Code Studies: Thriving Otherwise." *Progress in Human Geography* (2020). https://doi.org/10.1177/0309132519899733.

Emerson, R., R. I. Fretz, and L. L. Shaw. *Writing Ethnographic Fieldnotes* (Chicago Guides). Chicago: University of Chicago Press, 1995.

Escobar, Arturo. *Designs for the Pluriverse: Radical Interdependence, Autonomy, and the Making of Worlds*. Durham, NC: Duke University Press, 2018.

Freud, Sigmund. *Beyond the Pleasure Principle*. London: Penguin UK, 2003.

Freud, Sigmund. *Moses and Monotheism*. New York: A. A. Knopf, 1939.

Gehl, Robert W. "The Archive and the Processor: The Internal Logic of Web 2.0." *New Media & Society* 13, no. 8 (2011): 1228–1244.

Gillespie, Tarleton. "The Politics of 'Platforms.'" *New Media & Society* 12, no. 3 (2010): 347–364.

Gordillo, Gastón R. *Rubble: The Afterlife of Destruction*. Durham, NC: Duke University Press, 2014.

Gordon, Avery F. *Ghostly Matters: Haunting and the Sociological Imagination*. Minneapolis: University of Minnesota Press, 2008.

Gottesman, Evan. *Cambodia after the Khmer Rouge: Inside the Politics of Nation Building*. New Haven, CT: Yale University Press, 2004.

Granville, Johanna. "Radio Free Europe and International Decision-Making during the Hungarian Crisis of 1956." *Historical Journal of Film, Radio and Television* 24, no. 4 (2004): 589–561.

Guillou, Anne Yvonne. "An Alternative Memory of the Khmer Rouge Genocide: The Dead of the Mass Graves and the Land Guardian Spirits [neak ta]." *South East Asia Research* 20, no. 2 (2012): 207–226.

Gupta, Akhil, and James Ferguson. "Discipline and Practice: The 'Field' as Site, Method, and Location in Anthropology." In *Anthropological Locations: Boundaries and Grounds of a Field Science*, ed. Akhil Gupta and James Ferguson, 1–46. Berkeley: University of California Press, 2006.

Hackett, Edward J., Olga Amsterdamska, Michael Lynch, and Judy Wajcman. *The Handbook of Science and Technology Studies*. 3rd ed. Cambridge, MA: MIT Press, 2008.

Halbwachs, Maurice. *On Collective Memory*. Chicago: University of Chicago Press, 1992.

Haraway, Donna. "Situated Knowledges: The Science Question in Feminism and the Privilege of Partial Perspective." *Feminist Studies* 14, no. 1 (1988): 575–599.

Harvey, David. "Memory, That Powerful Political Force." In *Space and the Memories of Violence: Landscapes of Erasure, Disappearance and Exception*, ed. Estela Schindel and Pamela Colombo, 244–253. London: Palgrave Macmillan UK, 2014.

Hecht, Gabrielle, ed. *Entangled Geographies: Empire and Technopolitics in the Global Cold War*. Cambridge, MA: MIT Press, 2011.

Heder, Steve, and Judy Ledgerwood. "Introduction." In *Propaganda, Politics and Violence in Cambodia*, ed. Steve Heder and Judy Ledgerwood, 1–47. Armonk, NY: M. E. Sharpe, 1996.

Hedges, Inez. *World Cinema and Cultural Memory*. Houndmills: Palgrave Macmillan, 2015.

Hinrichsen, Lisa. "Trauma Studies and the Literature of the US South." *Literature Compass* 10, no. 8 (2013): 605–617.

Hinton, Alexander Laban. *The Justice Façade: Trials of Transition in Cambodia*. Oxford: Oxford University Press, 2018.

Hinton, Alexander Laban. *Why Did They Kill? Cambodia in the Shadow of Genocide*. Berkeley: University of California Press, 2005.

Hirsch, Marianne. "The Generation of Postmemory." *Poetics Today* 29, no. 1 (2008): 103–128.

Hochschild, Arlie. "Foreword: Invisible Labor, Inaudible Voice." In *Invisible Labor*, ed. Marion Crain, Winifred Poster, and Miriam Cherry, xi–xiv. Oakland: University of California Press, 2016.

Hooper, Andy. "Green Computing." *Communication of the ACM* 51, no. 10 (2008): 11–13.

Hu, Tung-Hui. *A Prehistory of the Cloud*. Cambridge, MA: MIT Press, 2015.

Huang, Yasheng. *Capitalism with Chinese Characteristics: Entrepreneurship and the State*. Cambridge: Cambridge University Press, 2008.

Huffman, Franklin E. *Cambodian System of Writing and Beginning Reader with Drills and Glossary*. New Haven, CT: Yale University Press, 1970.

Hughes, Caroline, and Netra Eng. 2019. "Facebook, Contestation and Poor People's Politics: Spanning the Urban–Rural Divide in Cambodia?" *Journal of Contemporary Asia* 49, no. 3 (2019): 365–388.

Ingawanij, May Adadol. "Itinerant Cinematic Practices in and around Thailand during the Cold War." *Southeast of Now: Directions in Contemporary and Modern Art in Asia* 2, no. 2 (2018): 9–41.

Irani, Lilly. *Chasing Innovation: Making Entrepreneurial Citizens in Modern India*. Princeton, NJ: Princeton University Press, 2019.

Irani, Lilly, Janet Vertisi, Paul Dourish, Kavita Philip, and Rebecca Grinter. "Postcolonial Computing: A Lens on Design and Development." In *Proceedings of the SIGCHI Conference on Human Factors in Computing Systems 2010*, 1311–1320. New York: ACM, 2010.

Jack, Margaret, and Seyram Avle. "A Feminist Geopolitics of Technology." *Global Perspectives* 2, no. 1 (2021): 24398.

Jack, Margaret, Jay Chen, and Steven J. Jackson. "Infrastructure as Creative Action: Online Buying, Selling, and Delivery in Phnom Penh." In *Proceedings of the 2017 CHI Conference on Human Factors in Computing Systems*. New York: ACM, 2017.

Jack, Margaret, Pang Sovannaroth, and Nicola Dell. "'Privacy Is Not a Concept, but a Way of Dealing with Life': Localization of Transnational Technology Platforms and Liminal Privacy Practices in Cambodia." *Computer Supported Cooperative Work*, November 2019.

Jackson, Steven J. "Rethinking Repair." In *Media Technologies: Essays on Communication, Materiality, and Society*, ed. Tarleton Gillespie, Pablo J. Boczkowski, and Kirsten A. Foot, 221–239. Cambridge, MA: MIT Press, 2014.

Jelin, Elizabeth. *State Repression and the Labors of Memory*. Minneapolis: University of Minnesota Press, 2003.

Johnson, Andrew Alan. *Ghosts of the New City: Spirits, Urbanity, and the Ruins of Progress in Chiang Mai*. Honolulu: University of Hawai'i Press, 2014.

Kelly, Caleb. *Cracked Media: The Sound of Malfunction*. Cambridge, MA: MIT Press, 2009.

Kiernan, Ben. *The Pol Pot Regime: Race, Power, and Genocide in Cambodia under the Khmer Rouge, 1975–79*. New Haven, CT: Yale University Press, 2002.

Kittler, Friedrich A. *Gramophone, Film, Typewriter*. Stanford, CA: Stanford University Press, 1999.

Kleinman, Arthur, and Joan Kleinman. "The Appeal of Experience; The Dismay of Images: Cultural Appropriations of Suffering in Our Times." *Daedalus* 125, no. 1 (1996): 1–23.

Krapp, Peter. *Noise Channels: Glitch and Error in Digital Culture*. Minneapolis: University of Minnesota Press, 2011.

Kusno, Abidin. *After the New Order: Space, Politics, and Jakarta*. Honolulu: University of Hawai'i Press, 2013.

Kwon, Heonik. *After the Massacre: Commemoration and Consolation in Ha My and My Lai*. Berkeley: University of California Press, 2006.

Lackaff, Derek, and William J. Moner. "Local Languages, Global Networks: Mobile Design for Minority Language Users." In *Proceedings of the 34th ACM International Conference on the Design of Communication*. New York: ACM, 2016.

Larkin, Brian. "The Politics and Poetics of Infrastructure." *Annual Review of Anthropology* 42 (2013): 327–343.

Larkin, Brian. *Signal and Noise: Media, Infrastructure, and Urban Culture in Nigeria*. Durham, NC: Duke University Press, 2008.

Lee, Doreen. *Activist Archives: Youth Culture and the Political Past in Indonesia*. Durham, NC: Duke University Press, 2016.

Lee, Sangjoon. "Creating an Anti-Communist Motion Picture Producers' Network in Asia: The Asia Foundation, Asia Pictures, and the Korean Motion Picture Cultural Association." *Historical Journal of Film, Radio and Television* 37, no. 3 (2017): 517–538.

Leos, Raymond. "Economic Ties with U.S. Cut in 1963; Relations in 1965." *Cambodia Daily*, October 11, 2016. https://english.cambodiadaily.com/news/letter-editor-economic-ties-u-s-cut-1963-relations-1965-119123/.

Lindtner, Silvia M. *Prototype Nation: China and the Contested Promise of Innovation*. Princeton, NJ: Princeton University Press, 2020.

Little, Harriet Fitch, and Vandy Muong. "Relics of Cambodia's Cinematic Golden Age." *Phnom Penh Post*, December 19, 2014. https://www.phnompenhpost.com/post-weekend/relics-cambodias-cinematic-golden-age.

Ly, Boreth. *Traces of Trauma: Cambodian Visual Culture and National Identity in the Aftermath of Genocide*. Honolulu: University of Hawai'i Press, 2020.

Marcus, George E. "Multi-Sited Ethnography: Five or Six Things I Know about It Now." In *Multi-Sited Ethnography: Theory, Practice and Locality in Contemporary Research*, ed. Mark-Anthony Falzon, 24–40. Farnham: Ashgate, 2012.

Marston, John. "Cambodian News Media in the UNTAC Period." In *Propaganda, Politics and Violence in Cambodia: Democratic Transition under United Nations Peace-Keeping*, ed. Steve Heder and Judy Ledgerwood, 208–242. Armonk, NY: M. E. Sharpe, 1996.

Mattern, Shannon. "Scaffolding, Hard and Soft: Media Infrastructures as Critical and Generative Structures." In *The Routledge Companion to Media Studies and Digital Humanities*, ed. Jentery Sayers. New York: Routledge, 2018.

McLuhan, Marshall. *Understanding Media: The Extensions of Man*. 1964. Reprint, Cambridge, MA: MIT Press, 1994.

Mech, Dara, and Ananth Baliga. "Death of Democracy: CNRP Dissolved by Supreme Court Ruling." *Phnom Penh Post*, November 17, 2017. https://www.phnompenhpost.com/national-post-depth-politics/death-democracy-cnrp-dissolved-supreme-court-ruling.

Mekong Strategic Partners and Raintree. "Cambodian Startup Ecosystem Report 2018." 2018. https://www.raintreecambodia.com/research.

Menkman, Rosa. "Glitch Studies Manifesto." In *Video Vortex Reader II: Moving Images beyond YouTube*, ed. Geert Lovink and R. Miles, 336–347. Amsterdam: Institute of Network Cultures, 2011.

Miller, Daniel. *Tales from Facebook*. Cambridge: Polity, 2011.

Mrázek, Rudolf. *Engineers of Happy Land: Technology and Nationalism in a Colony*. Princeton, NJ: Princeton University Press, 2018.

Muan, Ingrid. "Citing Angkor: The 'Cambodian Arts' in the Age of Restoration, 1918–2000." Dissertation, Columbia University, 2002.

Muan, Ingrid. "Playing with Powers: The Politics of Art in Newly Independent Cambodia." *UDAYA, Journal of Khmer Studies* 6 (2005): 41–56.

Muan, Ingrid, and Ly Daravuth. "A Survey of Film in Cambodia." In *Film in South East Asia: Views from the Region*, ed. David Hanan, 93–106. Manila: SEAPAVAA in association with the Vietnam Film Institute and the National Screen and Sound Archive of Australia, 2001.

Muong, Vandy, and Harriet Fitch Little. "The Man Who Painted Cambodian Cinema's 'Golden Age.'" *Phnom Penh Post*, March 18, 2016. https://www.phnompenhpost.com/post-weekend/man-who-painted-cambodian-cinemas-golden-age.

Nelson, Diane M. *Reckoning: The Ends of War in Guatemala*. Durham, NC: Duke University Press, 2009.

Nguyen, Lilly U. "Infrastructural Action in Vietnam: Inverting the Techno-Politics of Hacking in the Global South." *New Media & Society* 18, no. 4 (2016): 637–652.

Nora, Pierre, and Lawrence D. Kritzman, eds. *Realms of Memory: The Construction of the French Past*. New York: Columbia University Press, 1996.

Norén-Nilsson, Astrid. *Cambodia's Second Kingdom: Nation, Imagination, and Democracy*. Ithaca, NY: Cornell University Press, 2018.

Nunes, M. "Error, Noise, and Potential: The Outside of Purpose." In *Error: Glitch, Noise, and Jam in New Media Cultures*, 3–26. New York: Continuum, 2011.

Oakman, Daniel. *Facing Asia: A History of the Colombo Plan*. Canberra: ANU Press, 2010.

Olsen, Jon Berndt. *Tailoring Truth: Politicizing the Past and Negotiating Memory in East Germany, 1945–1990*. Oxford: Berghahn Books, 2015.

Ong, Aihwa, and Stephen J. Collier, eds. *Global Assemblages: Technology, Politics, and Ethics as Anthropological Problems*. Hoboken: John Wiley & Sons, 2008.

Osborne, Milton E. *Sihanouk: Prince of Light, Prince of Darkness*. Honolulu: University of Hawai'i Press, 1994.

Parikka, Jussi. "Mapping Noise: Techniques and Tactics of Irregularities, Interception, and Disturbance." In *Media Archaeology: Approaches, Applications, and Implications*, ed. Erkki Huhtamo and Jussi Parikka, 256–277. Berkeley: University of California Press, 2011.

Parikka, Jussi. *What Is Media Archaeology?* New York: John Wiley & Sons, 2013.

Parks, Lisa, and Nicole Starosielski, eds. *Signal Traffic: Critical Studies of Media Infrastructures*. Urbana: University of Illinois Press, 2015.

Phnom Penh Post staff. "Sokha Arrested for Treason, Is Accused of Colluding with US to Topple the Government." *Phnom Penh Post*, September 4, 2017. www.phnompenhpost.com/national/sokha-arrested-treason-accused-colluding-us-topple-government.

Pipek, Volkmar, and Volker Wulf. "Infrastructuring: Toward an Integrated Perspective on the Design and Use of Information Technology." *Journal of the Association for Information Systems* 10, no. 5 (2009): 447–473.

Poster, Winifred R. "Who's on the Line? Indian Call Center Agents Pose as Americans for US-Outsourced Firms." *Industrial Relations: A Journal of Economy and Society* 46, no. 2 (2007): 271–304.

Punzalan, Ricardo L., and Michelle Caswell. "Critical Directions for Archival Approaches to Social Justice." *The Library Quarterly* 86, no. 1 (2016): 25–42.

Quigley, Cassie, Alandeom W. Oliviera, Alastair Curry, and Gayle Buck. "Issues and Techniques in Translating Scientific Terms from English to Khmer for a University-Level Text in Cambodia." *Language, Culture and Curriculum* 24, no. 2 (2011): 159–177.

Radio Free Asia staff. "Cambodia to Monitor, Control Online News Ahead of Upcoming Ballot." *Radio Free Asia*, June 4, 2018. https://www.rfa.org/english/news/cambodia/news-06042018162755.html.

Radstone, Susannah. "Trauma Studies: Contexts, Politics, Ethics." In *Other People's Pain: Narratives of Trauma and the Question of Ethics*, ed. Martin Modlinger and Philipp Sonntag, 63–90. Oxford: Peter Lang, 2011.

Reddick, James, and Rinith Taing. "After the Khmer Rouge, the 'Most Beautiful' Voice Called Holdouts Home." *Phnom Penh Post*, May 26, 2017. https://www.phnompenhpost.com/post-depth-post-life-arts-culture/after-khmer-rouge-most-beautiful-voice-called-holdouts-home.

Reporters without Borders. "Cambodia: The Independent Press in Ruins." 2017. https://rsf.org/sites/default/files/cambodia_report_eng.pdf.

Reuters. "Cambodia Blocks Some Independent News Media Sites: Rights Group." July 27, 2018. https://www.reuters.com/article/us-cambodia-election-censorship/cambodia-blocks-some-independent-news-media-sites-rights-group-idUSKBN1KH29Q.

Reuters. "Cambodia Daily Shuts with 'Dictatorship' Parting Shot at Prime Minister Hun Sen." *The Guardian*, September 3, 2017. https://www.theguardian.com/world/2017/sep/04/cambodia-daily-shuts-with-dictatorship-parting-shot-at-prime-minister-hun-sen.

Richards, Thomas. *The Imperial Archive: Knowledge and the Fantasy of Empire*. London: Verso, 1993.

Rouse, Sarah. "South Vietnam's Film Legacy." *Historical Journal of Film, Radio and Television* 6, no. 2 (1986): 211–222.

Rust, William J. *Eisenhower and Cambodia: Diplomacy, Covert Action, and the Origins of the Second Indochina War*. Lexington: University Press of Kentucky, 2016.

"Safeguarding Peace: Cambodia's Constitutional Challenge." Special issue of *Accord: An International Review of Peace Initiatives*, no. 5 (November 1998).

Sambasivan, N., Ed Cutrell, Kentaro Toyama, and Bonni Nardi. "Intermediated Technology Use in Developing Communities." In *Proceedings of the SIGCHI Conference on Human Factors in Computing*, 2583–2592. New York: ACM, 2010.

Sarkar, Bhaskar, and Janet Walker. "Introduction: Moving Testimonies." In *Documentary Testimonies: Global Archives of Suffering*, ed. Bhaskar Sarkar and Janet Walker, 1–34. New York: Routledge, 2009.

Schlund-Vials, Cathy J. *War, Genocide, and Justice: Cambodian American Memory Work*. Minneapolis: University of Minnesota Press, 2012.

Schwenkel, Christina. *The American War in Contemporary Vietnam: Transnational Remembrance and Representation*. Bloomington: Indiana University Press, 2009.

Schwenkel, Christina. "Haunted Infrastructure: Religious Ruins and Urban Obstruction in Vietnam." *City & Society* 29, no. 3 (2017): 413–434.

Schwenkel, Christina. "Recombinant History: Transnational Practices of Memory and Knowledge Production in Contemporary Vietnam." *Cultural Anthropology* 21, no. 1 (2006): 3–30.

Sereypagna, Pen. "New Khmer Architecture: Modern Movement in Cambodia between 1953 and 1970." *Docomomo Journal* 57 (2017): 12–20.

Shawcross, William. *The Quality of Mercy: Cambodia, Holocaust and Modern Conscience.* London: Fontana/Collins, 1984.

Shulevitz, Judith. "The Science of Suffering." *The New Republic*, November 16, 2014. https://newrepublic.com/article/120144/trauma-genetic-scientists-say-parents-are-passing-ptsd-kids.

Simmel, George. "Two Essays (The Handle, The Ruin)." *The Hudson Review* 11, no. 3 (1958): 371–385.

Sinpeng, Aim, and Ross Tapsell, eds. *From Grassroots Activism to Disinformation: Social Media in Southeast Asia*. Singapore: ISEAS-Yusof Ishak Institute, 2020.

Slocomb, Margaret. *The People's Republic of Kampuchea, 1979–1989: The Revolution after Pol Pot*. Chiang Mai: Silkworm Books, 2003.

Sokhean, Ben. "Hun Sen: Peace Brought by Khmers, Not 'Foreign Hands.'" *Phnom Penh Post*, March 26, 2018. https://www.phnompenhpost.com/national/hun-sen-peace-brought-khmers-not-foreign-hands.

Star, Susan Leigh. "The Ethnography of Infrastructure." *American Behavioral Scientist* 43, no. 3 (1999): 377–391.

Star, Susan Leigh, and Geoffrey C. Bowker. "Enacting Silence: Residual Categories as a Challenge for Ethics, Information Systems, and Communication." *Ethics and Information Technology* 9, no. 4 (2007): 273–280.

Star, Susan Leigh, and Karen Ruhleder. "Steps toward an Ecology of Infrastructure: Design and Access for Large Information Spaces." *Information Systems Research* 7, no. 1 (1996): 111–134.

Star, Susan Leigh, and Anselm Strauss. "Layers of Silence, Arenas of Voice: The Ecology of Visible and Invisible Work." *Computer Supported Cooperative Work (CSCW)* 8, nos. 1–2 (1999): 9–30.

Steinberg, David J. *Cambodia: Its People, Its Society, Its Culture*. New Haven, CT: HRAF Press, 1957.

Steyerl, Hito. "In Defense of the Poor Image." *e-flux journal* 10, no. 11 (2009).

Stoler, Ann Laura. *Along the Archival Grain: Epistemic Anxieties and Colonial Common Sense*. Princeton, NJ: Princeton University Press, 2010.

Stoler, Ann Laura. "Imperial Debris: Reflections on Ruins and Ruination." *Cultural Anthropology* 23, no. 2 (2008): 191–219.

Strangio, Sebastian. *Hun Sen's Cambodia*. New Haven, CT: Yale University Press, 2014.

Strassler, Karen. *Refracted Visions: Popular Photography and National Modernity in Java*. Durham, NC: Duke University Press, 2010.

Strauss, Anselm. *Continual Permutations of Action*. New York: Walter de Gruyter, 1993.

Strauss, Anselm. *Qualitative Analysis for Social Scientists*. Cambridge: Cambridge University Press, 1987.

Strauss, Anselm, and Juliet Corbin, *Basics of Qualitative Research: Techniques and Procedures for Developing Grounded Theory*. Thousand Oaks, CA: Sage, 1998.

Swan, Christian. "Voice of America, a Radio Heard in Secret." *Christian Science Monitor*, February 7, 1980.

Thompson, Ashley. "Forgetting to Remember, Again: On Curatorial Practice and 'Cambodian Art' in the Wake of Genocide." *diacritics* 41, no. 2 (2013): 82–109.

Thompson, Ashley, and Stephen Murphy. "Cambodia Is Turning the Tide on Looted Statues, but Some Things Cannot Be Returned." *The Guardian*, February 21, 2021. https://www.theguardian.com/commentisfree/2021/feb/13/cambodia-looted-statues-ancient-khmer-objects.

Todd, Zoe. "An Indigenous Feminist's Take on the Ontological Turn: 'Ontology' Is Just Another Word for Colonialism." *Journal of Historical Sociology* 29, no. 1 (2016): 4–22.

Tsing, Anna Lowenhaupt. *The Mushroom at the End of the World: On the Possibility of Life in Capitalist Ruins*. Princeton, NJ: Princeton University Press, 2015.

Tufekci, Zeynep. *Twitter and Tear Gas: The Power and Fragility of Networked Protest*. New Haven, CT: Yale University Press, 2017.

Turner, Fred. *The Democratic Surround: Multimedia and American Liberalism from World War II to the Psychedelic Sixties*. Chicago: University of Chicago Press, 2013.

Uk, Krisna. *Salvage: Cultural Resilience among the Jorai of Northeast Cambodia*. Ithaca, NY: Cornell University Press, 2016.

Um, Khatharya. *From the Land of Shadows*. New York: New York University Press, 2015.

Updegraff, John A., Roxane Cohen Silver, and E. Alison Holman. "Searching for and Finding Meaning in Collective Trauma: Results from a National Longitudinal Study of the 9/11 Terrorist Attacks." *Journal of Personality and Social Psychology* 95, no. 3 (2008): 709–722.

Van der Kolk, Bessel A. *The Body Keeps the Score: Brain, Mind, and Body in the Healing of Trauma.* New York: Penguin Books, 2015.

Van Dijck, José. *Mediated Memories in the Digital Age.* Stanford, CA: Stanford University Press, 2007.

Vichheika, Kann. "RFA Journalists Accused of Treason Released on Bail after Nine Months behind Bars." *VOA*, August 21, 2018. https://www.voacambodia.com/a/rfa-journalists-accused-of-treason-released-on-bail-after-nine-months-behind-bars/4537923.html.

Vong, Mun, and Kimhean Hok. "Facebooking: Youth's Everyday Politics in Cambodia." *South East Asia Research* 26, no. 3 (2018): 219–234.

Vong, Mun, and Aim Sinpeng. "Cambodia: From Democratization of Information to Disinformation." In *From Grassroots Activism to Disinformation: Social Media in Southeast Asia*, ed. Aim Sinpeng and Ross Tapsell. Singapore: ISEAS, 2020.

Vuth, Lyno. "Knowledge Sharing and Learning Together: Alternative Art Engagement from Stiev Selapak and Sa Sa Art Projects (En/FR/KR)." *UDAYA, Journal of Khmer Studies* 12 (2015): 253–302.

Weltevrede, Esther, Anne Helmond, and Carolin Gerlitz. "The Politics of Real-Time: A Device Perspective on Social Media Platforms and Search Engines." *Theory, Culture & Society* 31, no. 6 (2014): 125–150.

Williams, Raymond, and Michael Orrom. *Preface to Film.* N.p.: London Film Drama, 1954.

Willis, P., and M. Trondman. "Manifesto for Ethnography." *Ethnography* 1, no. 1 (2000): 5–16.

Wolfinger, Nicholas. "On Writing Fieldnotes." *Qualitative Research* 2, no. 1 (2002): 85–93.

Wolford, W. "The Difference Ethnography Can Make: Understanding Social Mobilization and Development in the Brazilian Northeast." *Qualitative Sociology* 29 (2006): 335–352.

Wyche, Susan, Tawanna R. Dillahunt, Nightingal Simiyu, and Sharon Alaka. "If God Gives Me the Chance I Will Design My Own Phone: Exploring Mobile Phone Repair and Postcolonial Approaches to Design in Rural Kenya." In *Proceedings of the 2015 ACM International Joint Conference on Pervasive and Ubiquitous Computing*, 463–473. New York: ACM, 2015.

Young, A., Jr. "Coming Out from under the Ethnographic Interview." Workshop on Interdisciplinary Standards for Systematic Qualitative Research, 2007. http://www.wjh.harvard.edu.

Zaugg, Isabeelle. "Digitizing Ethiopic: Coding for Linguistic Continuity in the Face of Digital Extinction." Dissertation, American University, 2017.

Zhou, Mei. *Radio UNTAC of Cambodia: Winning Ears, Hearts, and Minds*. Bangkok: White Lotus, 1994.

Zuboff, Shoshana. *The Age of Surveillance Capitalism: The Fight for a Human Future at the New Frontier of Power*. London: Profile Books, 2019.

DOCUMENTARIES AND FILMS

Ayuth, Mao, and Lim Kvang Ngoc, dirs. *The Season of the Palm Flowers* (Rodau Pka Tnaout). 1980, 24 min.

Chou, Davy, dir. *Golden Slumbers*. France/Cambodia: Icarus Films, 2012. 96 min.

Mitterand, Frederic, dir. *Norodom Sihanouk, King and Film-Maker*. France: CasaDei Productions, 1997. 65 min.

Panh, Rithy, dir. *The Land of Wandering Souls*. France/Cambodia: INA, 2000. 100 min.

Panh, Rithy, dir. *The Missing Picture*. France/Cambodia: Les Acacias, 2013. 92 min.

Pirozzi, John, dir. *Don't Think I've Forgotten*. USA/Cambodia: Argot Pictures, 2014. 106 min.

Sihanouk, Norodom, dir. *Rose of Bokor*. Cambodia: Khemara Pictures, 1969. 70 min.

INDEX

Amazing Cambodia, 21, 159–168, 169, 175, 180, 181, 182, 191, 217n35
Arab Spring, 188
Auriol, Vincent, 31
AutoCAD, 111, 126–128, 187, 217n37
Ayai roeurng, 55, 63

Baksbat, 145, 147, 157
Battambang, 65, 70–71, 76, 109–110, 164, 202n45, 204n81, 215n46
Batteries, 29, 50, 65, 90, 92, 180
Bird of Paradise (film), 43, 45
Boats (cinema), 34
Bophana Center, 21, 50, 59, 161, 170–172, 177–180, 181, 182, 188
  App-learning on Khmer Rouge History, 21, 177–180, 182, 187, 188, 190, 210n57, 218n3
Bou Vannarith, 55, 59–66, 78, 79, 96, 97, 206–207n3
Braangk, 165–166, 169

Cambodia Daily, 122, 212n4
Cambodian 2018 general election, 2, 32, 112, 122
Cambodian National Rescue Party (CNRP), xi, 2, 112, 122, 123
Cambodian People's Party (CPP), 81, 82, 88, 95, 96, 122, 123, 212n4, 215n42
Cambodia's Reunion Programme, 99
Cell phone towers, 169–170

Censorship law, 46–47
Chakrya Moa, 87
Chea Sopheap, 170
Chenla Theater, 126–128
Chet Chong Cham (I Want to Remember; film), 69, 70
Chilling effects, 133, 191
China, 12, 38, 39, 77, 97, 147, 191, 200n27
Chou En Lai, 39. *See also* Zhou Enlai
Chun, Wendy, 161, 168–169
Cinéstar, 1–2, 112–114
Coalition Government of Democratic Kampuchea (CGDK), 54
Cold War, 13, 17, 18, 19, 20, 28, 29, 31, 39, 45, 54, 81, 105, 118, 173, 181, 188, 189, 190, 199n19, 207n7
Collective trauma, 5
Confederate statues, 4
Corporate self-regulation, 192
Cultural centers, 34, 39, 47

Dap Chhuon, 38, 46, 188
Data centers, 168, 191, 193
Debt, 141–142
Decolonization, 13, 39
Department of Culture and Propaganda, 66
Derrida, Jacques, 113, 216n5, 221n12
Digital economy, 13
Digital library, Southeast Asia, 28

Disintegration noise, 21, 135–136, 149–150, 154–155, 219n22
Dith Pran, 49–50
Dy Saveth, 140

Eden Cinema, 114
Eisenhower, Dwight D., 33
El-Tahri, Jihan, 5–6
Enchanted Forest, The (film), 44
Enduring ephemeral, 161, 169, 180–181
Ethnography, 11, 14, 15–16, 18
Extraordinary Chambers in the Courts of Cambodia (ECCC), 146

Facebook, 3, 13, 16, 123–124, 150–151, 156, 159, 161, 162, 169, 175–177, 191, 197n48, 216n58, 217n23, 221n5, 222n26
FARK, 43
Fiber-optic cable, 87, 102, 104, 105, 186, 188
Foreign Broadcast Information Service (FBIS), 58, 87, 207n7
Foreign interference, 13, 26, 79, 86, 98, 187
French Indochina, xi, 19, 29, 30
FUNCINPEC, xi, 53, 54, 81–82, 88, 95, 96, 97, 212–213n4, 215n42

Geopolitics of technology, 7, 14, 58, 79, 190
"Ghostly matters," 10. *See also* Hauntings
Glitches, 135, 148, 149, 155
Glitch politics, 14
Golden age, 1, 26, 44, 47, 69, 138
Golden Violence (film), 135, 151
Google Maps, 111, 117
Gordillo, Gaston, 124–125. *See also* Rubble
Gordon, Avery, 10, 113, 125, 133
Guillou, Anne, 11, 83, 146, 219n19, 220n36

Happiness workers, 8
Hauntings, 9, 10, 11, 21, 113, 125, 186
Hecht, Gabrielle, 58, 207n6
Heder, Steve, 82, 87, 88, 89, 97
Hemakcheat cinema, 117–119, 189
Heyman, Jeffery, 87, 89, 90, 96
Huber, Alex, 92
Hun Sen, xi, 10, 53, 66, 81, 82, 96, 97, 214n40

IBC, 99, 212–213n4
Imperial archive, 9, 160, 173–175, 180, 181
Inequality, 2, 9, 11, 19, 26, 79, 85, 94, 103, 104, 106, 118, 133, 134, 156, 176, 189, 220n40
Information and communication technologies for development (ICTD), 174–175
Information halls, 35
Infrastructural action, 7
Infrastructural methodology, 26
Infrastructural restitution, v, 1–3, 5–10, 14, 19–21, 58, 65, 78–79, 98, 106, 111, 125, 133, 135, 148, 156, 182–183, 185–189, 192
Infrastructuring, 7
Izzi, Monique, 25

Jorai, 147
Joy of Life, The (film), 44

Kamala, Loak, 59, 76–77, 78, 79
Kamtech, 10
Ka Toy, 59, 70–76, 78, 79, 99, 100, 187, 210n57, 215n47
Kem Sokha, 112–113, 122
Keo Chanda, 61
Khmer Republic, 49
Khmer Serei, 38, 40, 46
Kim Nova, 135, 151, 153
Kim Son cinema, 115, 117
King Kong (film), 153

Kirirom, 42
Klein, Norbert, 87, 100–101
Kon Len Knyom, 129, 131–132
Koompi, 176
KPNLF (Khmer People's National Liberation Front), 53–54, 62–63, 64, 78, 88

Land of Wandering Souls, The (film), 102–104
Ledgerwood, Judy, 82, 83, 87, 88, 97
"Life of America" photo competition, 45
Lindtner, Silvia, 8, 11–12
Little Prince, The (film), 44
Lkhoan ayai, 42
Lkhoan niyeay, 42, 63
Lon Nol, xi, 10, 44, 49, 69, 71, 102, 171, 212n4
Loudspeakers, 29, 32, 38, 57, 62, 65–66, 78, 90, 209n36
Lux Cinema, 120–121, 210–211n57
Ly Bun Yim, 45, 151

Maek Cha cinema, 109–110
Mao Ayuth, 59, 66–70, 73, 78, 79, 100, 210n51
Marston, John, 95, 99, 208–209n31, 209n36, 212–213n4, 215n42
Materiality, 5, 6–7, 20, 21, 26, 27, 55, 56, 104, 137, 138, 149, 154, 186
May Ebihara, 28, 36, 37, 202n57
Media ruins, 21, 109, 111, 114, 118, 124, 125, 129, 132–133, 185, 186
Mei, Zhou, 87
Memory sickness, 10
Ministry of Information, 26, 30, 32, 33, 41, 43, 46, 59, 67, 123, 201–202n29, 206–207n3
Ministry of Posts and Telecommunications, 100, 102, 123
Missing Picture, The (documentary), 53, 170, 177

Mobile cinema, and cinecars, 26, 28, 29, 34, 44, 57, 73, 199n19
Moeun Chhay, 109–110
Motherland Appeal (Neaty Somleng Ompeaveneav Robos Meataphum), 58, 63–64
Muan, Ingrid, 28
Muouy Meun Allay, 137–141

National Archives of Cambodia, 28, 58, 87, 162, 198n5, 217n35
National Museum of Cambodia, 28
Neak Poan Productions, 45
Neak ta, 146–147
Neeung Jew Kreu Fa (film), 138–139, 142, 154
New Khmer Architecture, 118, 162
Nguyen, Lily, 7
Nguyen Vu Cap, 61
Norodom Ranariddh, xi, 53, 97
Norodom Sihanouk, 2, 19–20, 25–26, 30–32, 35, 36–37, 38–44, 46–47, 49, 66, 67, 81–82, 188, 189, 193, 197n1, 198n4, 204n88, 205n101
Norodom Suramit, 32, 33

Office of Film, 30, 42
One Laptop per Child, 11, 174
Oudong, 160, 163–165, 167, 169, 180
Oun Chhin, 122

Pak, Steven, 83, 87, 94, 213n9
Paris Peace Accords, 20, 81, 214n40
People's Republic of Kampuchea (PRK), v, xi, 5, 20, 53–54, 55, 58, 61, 68, 80, 88, 89, 98, 100, 186, 188
Phka Rik Phka Ruy (Blooming Flower, Withering Flower; film), 43
Pol Pot, 4, 10
Positionality, 17
Postal service, 33
Post-traumatic stress disorder (PTSD), 144

Preah Sorya, 21, 135–136, 137–138, 149–150, 152, 153–157, 182, 187, 188, 190, 210n57, 218n3
"Preparing for Cambodian New Year" (radio program), 55–57, 79
Projectors, 34, 35, 43, 47, 75, 100, 161, 186, 201n41, 215n47
Public address systems, 31, 32, 37
Pulled candy, 163–164

Radio Cambodge, 30, 32, 33, 36, 39, 41, 42, 47
Radiodiffusion National Khmère, 36. *See also* Radio Cambodge
Radio Free Asia, 123
Radio programming, 33, 49, 61, 90, 91, 99, 104, 212–213n4
Radio Saigon, 29, 30
Radio technology course, offered by PRK, 61
Radio UNTAC, 89–99
Radio waves, 26, 27, 39, 40–41, 42, 47, 62, 189
Réalités cambodgiennes (magazine), 28, 41, 42, 46, 129, 198n4
Receivers, radio, 31, 32, 36, 78, 186, 200n27, 202–203n57
Recombinant history, 10–11
Reconstruction, 2, 4, 5, 20, 26, 78, 106, 186, 214n40
REI Foundation, 161, 170, 181
Relationality, 20, 26, 27, 186
Repair, 58, 70, 133, 190, 193
Residual categories, 8
Restitution, 5–6
Richards, Thomas, 173, 221n4, 221n14. *See also* Imperial archive
Rithy Panh, xi, 2, 49
Rodau Pka Tnaout (*The Season of the Palm Flowers*; film), 58, 69, 210n51
Rose of Bokor (film), 25, 28, 30, 44
Ros Sereysothea, 141

Roung Kon, 1–2, 21, 111–121, 125–134, 187, 188, 189, 190, 220n41
Rubble, 21, 113, 124–125
Ruins, 21, 113

Sam Oeun, 135, 151, 153, 220n38
Sangkum Reastr Niyum, 2, 13, 19, 26, 32, 42, 118, 189, 204n89, 206n3 (chap. 2)
Sar Kassora, 140
Sa Sa Art Projects, 55–56, 132
Schanberg, Sydney, 49–50
Schwenkel, Christine, 10–11, 125
Shadow over Angkor (film), 38, 44
Siem Reap, 34, 38, 89, 96, 101, 138, 178
Sinn Sisamouth, 141
Sisowath Monireth, 29–30
Sisowath Monivong, 29
Sobennavong (film), 43
Sok, Loak, 109
Som Sam Al, 43, 44, 66, 193
Southeast Asia digital library, 28
Specters, 113
Star, Susan Leigh, 8–9, 58, 187
Steinberg, David, 35, 36, 37–38, 199n14, 202–203n57
"Structure of feeling," 113
Sun Bun Ly, 45

Tap Songva, 140
Technopolitics, 58
Testimonies, 51
Thai Film Archive, 135, 138
Theravada Buddhism, 32, 125, 146
Third Cinema, 45
Transcultural Psychosocial Organization, 51, 145
Transmitters, radio, 26, 32, 33, 41, 47, 57, 60, 62, 65, 78, 89, 90, 186, 189
Troung Kok cinema, 115–116
Twilight (film), 44, 204n88

United Nations Transitional Authority
   of Cambodia (UNTAC), 20, 70, 77,
   81–83, 85–86, 88, 96–98, 104–105,
   123, 188, 189, 212n1, 214n38
   Information/Education Division, 82,
      88, 89, 92, 215n42
   media charter, 88
United States Information Service
   (USIS), 13, 19, 25–26, 28, 29, 31–36,
   40, 43, 45–47, 86, 189, 192, 193,
   199n19, 201n39, 202n47

Van Molyvann, 44, 120
Voice of America (VOA), 32, 49, 50, 89,
   122, 201n31
Voice of Democratic Kampuchea (radio
   station), 62
Voice of the Cambodian People, 60, 88
Voice of the Great National Union
   Front of Cambodia, 88
Voice of the Khmer People (radio
   station), 62, 64, 88, 213n6
Voice of the National Army of
   Democratic Kampuchea (radio
   station), 62

Wapatoa (blog), 176
Wi-Fi, 171, 180, 182
Williams, Raymond, 113

Yeang Sothearin, 122
Yvon Hem, 45, 100

Zhou Enlai, 39